Environmentalism—Pathways to Life for Humanity and Sustainability

RIVER PUBLISHERS SERIES IN SOCIAL, URBAN, ECONOMIC AND ENVIRONMENTAL SUSTAINABILITY

The River Series on Social, Urban, Economic and Environmental Sustainability is a series of comprehensive academic and professional books which focus on the societal side of sustainability. The series focuses on topics ranging from theory to policy and real-life case studies and applications.

Books published in the series include research monographs, edited volumes, handbooks and textbooks. The books provide professionals, researchers, educators, and advanced students in the field with an invaluable insight into the latest research and developments.

Topics included in this series are as follows:-

- Climate Change Mitigation
- Renewable Energy Policy
- Urban sustainability
- Strategic environmental planning
- Environmental Systems Monitoring and Analysis
- Greening the World Economy
- Sustainable Development

For a list of other books in this series, visit www.riverpublishers.com

Environmentalism—Pathways to Life for Humanity and Sustainability

Medani P. Bhandari
Akamai University, USA
Sumy State University, Ukraine
Gandaki University, Nepal

River Publishers
Routledge
Taylor & Francis Group
NEW YORK AND LONDON

Published 2025 by River Publishers
River Publishers
Alsbjergvej 10, 9260 Gistrup, Denmark
www.riverpublishers.com

Distributed exclusively by Routledge
605 Third Avenue, New York, NY 10017, USA
4 Park Square, Milton Park, Abingdon, Oxon OX14 4RN

Environmentalism—Pathways to Life for Humanity and Sustainability / Medani P. Bhandari.

Routledge is an imprint of the Taylor & Francis Group, an informa business

ISBN 978-87-7004-812-5 (hardback)
ISBN 978-87-7004-821-7 (paperback)
ISBN 978-87-7004-814-9 (online)
ISBN 978-8-770-04813-2 (ebook master)

To my friend, wife, coauthor Mrs. Prajita Bhandari, without your help and support this book was impossible.

Together We Have Miles To Go

You have assumed a pivotal role akin to a true teacher, actively fostering an environmentally conscious atmosphere within your home, workplace, and community. Your impactful actions and exemplary behavior serve as an inspiration, motivating others to embrace sustainable practices and embrace a greener way of life. Your endeavors encompass a broad spectrum, from waste reduction to energy conservation and the advocacy for eco-friendly initiatives. Through these contributions, you actively create a better world for all. Your unwavering dedication to the stewardship of the environment shines as a guiding beacon, urging others to join the collective effort toward a healthier and more sustainable future. Amidst life's challenges, you provide me with a reassuring sense of comfort and a valuable lesson in resilience. For this, I extend my heartfelt gratitude to you.

Contents

Preface

Today's world stands at a crossroads, facing a planetary crisis that transcends borders and affects every aspect of life. Environmental degradation, climate change, social inequity, and unsustainable practices threaten not only the natural world but the very survival of humanity. Against this backdrop, *Environmentalism: Pathways to Life for Humanity and Sustainability* has been conceived. This book represents the culmination of over four decades of my research, activism, and practical efforts as an environmentalist, humanitarian, and social worker. Its purpose is to explore our multifaceted challenges while providing new insights, theoretical foundations, and practical pathways toward a sustainable and equitable future.

This work deeply reflects environmental transformation and sustainability theories and practices grounded in academic research and hands-on experience. I seek to bridge the gaps between knowledge and practice through these pages, highlighting the urgent need for innovative thinking and collective action. From the foundational concepts of sustainability and the Sustainable Development Goals to cutting-edge approaches in renewable energy, carbon sequestration, and biodiversity conservation, the book is designed to address the pressing need for systemic change in how we interact with the planet and each other.

At the heart of this work is recognizing that sustainability is not just an environmental issue but a social and economic imperative. Social equity, justice, and global cooperation are integral to achieving a balanced future. I examine how resource inequality, corporate responsibility, and behavioral change must be part of the solution to achieving environmental justice and creating a sustainable economy. Additionally, I address the pivotal role of transformative education and lifelong learning in empowering individuals and communities to contribute to sustainable development. We can only equip future generations to meet the challenges ahead through continuous learning, innovation, and knowledge sharing.

The book also delves into the need for practical innovations, such as regenerative agriculture, climate resilience strategies, and sustainable finance, that can pave the way for green economic transformation. I explore the potential of technological advancements, circular economy models, and global collaboration to drive the transition toward sustainability. These discussions are grounded in the understanding that economic transformation must align with environmental stewardship and social well-being to create a sustainable future.

This journey through the various dimensions of environmentalism is not only an intellectual endeavor but a call to action. I hope this book serves as a roadmap for policymakers, activists, educators, and future generations, providing a clear direction for addressing the intertwined crises of climate change, inequality, and unsustainability. Together, we can forge a path toward a future where humanity thrives in harmony with the natural world, and this work is my contribution to that shared mission.

This book explores a comprehensive range of environmentalism, sustainability, and development topics. It examines the theoretical foundations and practical innovations, offering a pathway to understanding the complexity of humanity's relationship with nature and the environment. It covers:

The historical evolution of environmentalism

Theories of sustainability and their applications

The concept of resource inequality and its impact on global development

Sustainable Development Goals (SDGs) and their role in shaping policy and action

Climate change mitigation strategies, including renewable energy, carbon sequestration, and biodiversity conservation

Social equity, environmental justice, and global justice frameworks

The role of technological innovation in transforming economies

The circular economy, green economy, and sustainable finance models

Behavioral change, mental health, and well-being as integral to environmentalism

Corporate responsibility and global cooperation

The Epistemology of Development Economy and Sustainability

I hope to contribute to the epistemology of the development economy by exploring how different ways of knowing shape our approach to sustainable development. By weaving together economic theories, environmental philosophies, and ethical considerations, the book provides insights into how we can rethink development models to align more closely with sustainability principles. It also critically examines existing paradigms, proposing new frameworks that place humanity and ecological balance at the core of economic practices.

Why This Book is Important

This book offers a timely and much-needed exploration of the pathways forward in a world grappling with unprecedented environmental challenges. It is not just another academic treatise on environmentalism but a call to action grounded in decades of practical experience. It presents a clear, actionable roadmap for change agents, policymakers, academics, and activists through theoretical insights and case studies.

New Knowledge Proposed

One of this book's key contributions is its focus on bridging the knowledge gap between traditional environmentalism and the innovative approaches required to address the planetary crisis. It introduces new ways of thinking about sustainability, drawing from multiple disciplines such as ecological economics, philosophy, anthropology, and technological innovation. By integrating these diverse perspectives, the book proposes a holistic framework for understanding and addressing environmental degradation, climate change, and social inequity.

Bridging the Knowledge Gap

This book also addresses a critical gap in the current literature—the need for an interdisciplinary, integrated approach to environmentalism and sustainability. Bringing together theoretical, practical, and empirical perspectives creates a bridge between academic knowledge and real-world application. It highlights the importance of lifelong learning, transformative education, and collaboration across sectors to achieve meaningful and lasting change.

Why Change Agents Should Care

This book is a vital resource for policymakers, activists, educators, and other change agents. It outlines the challenges we face and provides clear, actionable strategies for overcoming them. Whether working at the grassroots level or shaping national policy, this book equips you with the tools and insights necessary to drive transformative change. It emphasizes that environmentalism is not just about saving the planet—it's about creating a more just, equitable, and sustainable future for all humanity.

Acknowledging the Journey and Collective Contributions

As I have noted earlier, this book is the culmination of more than four decades of research, practical work, and personal commitment to environmentalism and humanitarian efforts. Over these many years, countless individuals, experts, colleagues, communities, and collaborators have contributed to my journey, each offering invaluable perspectives, knowledge, and support.

It is impossible to fully capture or cite all the individuals and experts whose ideas, discussions, and contributions have shaped this work. From grassroots activists to academic scholars, policymakers to fellow environmentalists, the insights presented in this book result from a collective effort—a shared mission toward a more just, sustainable, and harmonious world.

This book belongs to everyone who has been part of this long journey. I dedicate it to everyone who has worked tirelessly in pursuing environmental and humanitarian justice. While I may be the one writing these words, the knowledge and experiences within these pages reflect the collective wisdom of many.

A Shared Hope for the Future

I sincerely hope this book will offer new insights and innovative approaches to tackling the global environmental and humanitarian crises we face. The interconnected challenges of climate change, social inequality, biodiversity loss, and environmental degradation require collaborative, holistic solutions. Drawing from the experiences, ideas, and knowledge of a diverse community of thinkers and doers, we can carve out new pathways to sustainability, equity, and global cooperation.

I invite readers to engage with the ideas presented in this book and to join the ongoing global effort to transform our societies into ones that prioritize life—human life, planetary life, and the life of future generations. Only through collaboration, resilience, and innovation can we secure a sustainable and just world for all. To everyone who has shared this journey with me—thank you. Your contributions are embedded in every chapter, and this book is as much yours as it is mine.

Thank you
Prof. Medani P. Bhandari, PhD.

Acknowledgement

The process of writing *Environmentalism: Pathways to Life for Humanity and Sustainability* has been one of the most fulfilling and deeply personal endeavors of my life. This book reflects my academic and intellectual growth and my lifelong commitment to understanding the complex, interwoven relationships between humanity and the natural world. As I reach this moment of completion, I feel immense gratitude for the countless individuals, experiences, and moments that have shaped this work and the perspective it conveys.

This book is the culmination of my journey as an environmental activist, humanitarian, researcher, and knowledge producer. For decades, I have dedicated myself to exploring the urgent need for sustainability, justice, and equity—realities that transcend academic disciplines, national borders, and cultural boundaries. My path has been a relentless pursuit to grasp how we, as a global society, might learn to coexist in harmony with the natural world while preserving its integrity for future generations.

Along the way, I have been motivated and inspired by human and non-human beings, the living environment, and the social systems that shape our interactions with the earth. From my early days as a young activist to my work today in academic and policy circles, every encounter has taught me more about the profound interconnectedness of life. The people I've worked with—activists, scholars, policymakers, and community leaders—have generously shared their knowledge, hopes, and insights. These conversations have challenged and inspired me to continue advocating for change in the face of global challenges.

However, not only people have been my teachers. Nature itself has been a constant and humbling source of wisdom. The resilience of ecosystems in the face of destruction, the remarkable intelligence of wildlife, and the delicate balance of the earth's life-support systems have all deeply influenced my thinking. Nature's inherent intelligence and resilience remind us that the solutions we seek must honor the intrinsic value of all human and non-human life.

Each chapter of this book reflects my many lessons from these relationships—with people, nature, and the intricate web of life that binds us all. Whether tackling issues of biodiversity loss, climate change, or the broader struggles for social justice and equity, I have come to understand that the solutions we seek must be holistic and inclusive. We must strive to create a future where humanity and the natural world can coexist and thrive in mutual respect and harmony.

I present this book with deep gratitude to all those who have walked with me on this journey—whether human, environmental, or otherwise. It is a reflection of my own life's work and a collaborative and interwoven effort that speaks to the shared journey we all face. I hope this work offers new insights and inspires action, just as those who have guided and supported me have inspired my efforts.

First and foremost, I would like to extend my heartfelt thanks to my family, whose unwavering support and encouragement have been the cornerstone of my journey. Their belief in my vision has given me the strength to navigate the challenges encountered along the way. To my wife, Prajita, and daughter, Manaslu, your love, patience, and understanding during the long hours spent researching and writing have motivated me to understand the world and our shared purpose better. Your presence has made this journey deeply personal and warm.

I am also deeply indebted to my mentors, colleagues, and peers, who have influenced my thinking and contributed to our shared understanding of the intricate relationships between humanity, ethics, and social equity. Your insights, constructive feedback, and encouragement have been invaluable in refining my ideas and enhancing the clarity of this work. This book is a testament to our shared learning and the community we have built together.

Furthermore, I wish to acknowledge the countless individuals and communities I have encountered as an environmental conservationist and humanitarian. Your stories of resilience, compassion, and innovation have enriched my understanding of the core principles discussed in this book. I hope to honor your experiences by weaving them into the fabric of this narrative.

I sincerely thank my mentors and colleagues who have guided and challenged me throughout my career. Your insights, feedback, and collaborative spirit have greatly enriched my understanding and have been instrumental in developing this book. Special thanks to Prof. Steven R. Brechin, whose wisdom and encouragement have been invaluable, and to the academic institutions and communities, including Syracuse University, NY, Brandeis University, MA, USA, ITC, the Netherlands, and Tribhuvan University, Nepal, for the stimulating discussions and thoughtful critiques that have sharpened my perspective.

I am also grateful to the many scholars and researchers whose work has influenced and informed my own. The breadth and depth of knowledge in this field are vast, and the contributions of others have provided a rich tapestry of ideas upon which this book is built. To those whose writings and theories have inspired me, thank you for your intellectual generosity.

I extend my heartfelt gratitude to my dear friends, Mr. Kshitiz Raj Prasai, Rajan Adhikari, Dr. Medini Adhikari, Govinda Luitel, Binod Luitel and Prof. Bishwa Kalyan Parajuli. Your steadfast encouragement and insistence on continuing my research and writing have been a persistent driving force, bolstering my conviction in this meaningful endeavor.

I sincerely thank Prof. Mary Jo Bulbrook, Ph.D.; Prof. Ganesh Man Gurung, Ph.D.; Prof. Douglas Capogrossi, Ph.D.; Prof. Jacek Piotr Binda, Ph.D.; Dr. Ambika Adhikari, Ph.D.; Prof. Keshav Bhattarai, Ph.D.; Dr. Shvindina Hanna, Ph.D.; Dr. Aleksander Łukasz Sapiński; Prof. Durga Paudel, Ph.D.; Mr. Jay Bhandari, and others for their encouragement and support in my academic pursuits.

"Janani Janmabhoomischa Swargadapi Gariyasi," a timeless adage that resounds with wisdom, acknowledges the unparalleled significance of both mother and motherland. At the outset, I extend my heartfelt gratitude to my mother, Heema Devi, whose uniqueness is a testament to the essence of all mothers, present and future, and Mother Earth herself. The depth of a mother's devotion is immeasurable, transcending enumeration. A mother's love remains matchless—until we fathom the depth of a mother's boundless, selfless affection, comprehending the very nature of maternal care, it is impossible to fully grasp the profound love and nurturing that Mother Earth extends to us and all living beings. Just as a mother's love is irreplaceable, so too is the immeasurable contribution of mother nature—a contribution that sustains and supports life in all its forms.

I would also like to extend my sincere gratitude to the invaluable contributions of my beloved family members, who have been unwavering pillars of support throughout my conservation journey. My wife, Prajita, along with my son, Prameya, daughter, Manaslu, and daughter-in-law, Kelsey Bhandari, son-in-law Abhimanyu, and his parents, Uma and Mahesh Iyer, thank you for your insightful comments and unwavering support during challenging times. To our grandchildren, Neena and Neelok, who brought joy to my family, I am grateful for your presence. I would also like to remember my two other grandchildren, Prasmin and Aahana Bhandari, and I hope that one day you will be reading this and other books of mine with pride.

My brothers Krishna and Shova, along with my nephews Jagannatha, Asmita, and Muna, brothers Hari and Durga, and daughter Pratigya, have been steadfast companions on this journey. My sisters Kali, Bhakti, Radha, Bindu, and Sita and their families have also provided unwavering support.

Additionally, my uncle Bhim Prasad Bhandari and the late Goma Devi Bhandari, my siblings Govinda, Bishnu, Rudra, Bishma, and Sabita, together with their spouses and children, have been a constant source of strength and encouragement.

Prajita's parents, Dwarika Natha and Durga Devi, and her siblings Basudev, Ramesh, and Rishi, as well as Dhurba Acharya, Binod Ghimire, Ranjan Neupane, and Annapurna, Bijaya, Jaya, Ajaya, and their entire families have played an integral role in our endeavors. Without their unwavering support, my conservation journey would have faced interruptions or not have been possible. This sentiment applies universally—human beings thrive as social entities through mutual support.

A heartfelt and special acknowledgment is dedicated to my wife, Mrs. Prajita Bhandari, whose presence has created a tranquil space that enabled the completion of this book. Your perceptive insights and meticulous language editing have played an indispensable role in molding the outcome. You have woven new threads into our journey—our endeavors in conservation and humanitarianism—and your support has been my strength throughout the trials of my life. Words are inadequate to convey my gratitude, and it is worth noting that this book is not just my narrative but also a testament to the journey we are embarking upon together.

I want to give a special thanks to the River Publishers team for their support and assistance in publishing this book. And to the future generations, whose well-being and future I hold close to my heart, thank you for motivating me to act and work towards a sustainable and thriving world for you.

Last but not least, I want to express my gratitude to all the readers. Your interest and engagement with my work fuels my passion for environmental conservation and social empowerment.

With a heart full of appreciation,

Thank you!

Prof. Medani P. Bhandari, PhD.

Environmentalism—Pathways to Life for Humanity and Sustainability

Environmentalism—Pathways to Life for Humanity and Sustainability involves creating a harmonious balance between human development and the health of our planet. Humanity's future depends on recognizing the interconnectedness of our well-being with the natural world and adopting practices that ensure long-term sustainability. This pathway emphasizes the importance of fostering a deep sense of global citizenship, where each of us, as stewards of the Earth, plays a crucial role in preserving it for future generations.

Central to this is the concept of *sustainable development*, which seeks to meet the needs of the present without compromising the ability of future generations to meet their own. This involves integrating environmental, social, and economic considerations into all aspects of life, from policy-making to personal choices. By prioritizing renewable resources, reducing waste, and promoting equity, we can create systems that support human flourishing and ecological health and pave the way for a brighter, more sustainable future.

The pathway underscores the importance of ethical decision-making and compassionate action. By cultivating empathy and understanding, we can address social inequalities and environmental degradation in a way that benefits all. Ultimately, the pathways to life for humanity and sustainability are about creating a future where people and the planet thrive together, ensuring a legacy of resilience and prosperity.

"Environmentalism: Pathways to Life for Humanity and Sustainability" is the culmination of my more than four decades of work in environmental research, humanitarianism, and social activism. Throughout these years, I have witnessed the growing challenges of climate change, biodiversity loss, and social inequities that threaten our planet's and its people's sustainability. This book reflects my commitment to presenting a new direction for sustainability and humanity, emphasizing that with complete dedication to conserving nature and natural resources, fostering a healthy society or a thriving planet is possible.

The book delves into the concept of Sustainable Development, beginning with the importance of Balanced Growth, which prioritizes economic, social, and environmental harmony. It emphasizes Resource Efficiency, advocating for a shift from wasteful consumption to responsible management of natural resources. Integrating Renewable Energy is also crucial, as transitioning away from fossil fuels is necessary to reduce greenhouse gas emissions and promote long-term sustainability.

Another focal point is Climate Change Mitigation, which addresses the urgent need to reduce carbon emissions and slow global warming. Alongside this, Climate Adaptation strategies are explored, emphasizing the importance of resilience-building for communities, ecosystems, and economies in the face of inevitable climate impacts. The preservation of Biodiversity is highlighted as essential for maintaining the balance of ecosystems and ensuring the survival of countless species, including our own.

A significant section of the book is dedicated to Social Equity and Justice, emphasizing that environmental sustainability cannot be achieved without addressing inequalities that marginalize vulnerable communities. The interlinking of environmental justice with technological innovation is also explored, showcasing how new technologies can drive sustainable solutions while being mindful of ethical considerations.

The book champions the Circular Economy as a transformative approach to reducing waste and promoting resource reuse and recycling. This vision is supported by the need for Global Cooperation, an essential element for addressing transboundary environmental issues such as climate change, pollution, and deforestation. Moreover, Community Engagement is portrayed as a fundamental pillar for local and grassroots efforts in environmental conservation and sustainability initiatives.

As part of the broader economic vision, economic transformation is discussed, focusing on Job Creation in sectors that promote sustainability, such as renewable energy, conservation, and green infrastructure. The interdependence of Health and Well-being with environmental health is highlighted, addressing how Healthy Lifestyles and Mental Health are closely tied to a healthy environment, with pollution, environmental stress, and climate anxiety becoming growing concerns.

The book also elaborates on Ecosystem Services, the natural functions that support human life, and how preserving and restoring them is vital for long-term planetary health. Strategies for restoring imbalanced ecosystems are outlined, along with recognition of nature's intrinsic value and the economic and social benefits of preserving these services.

This book will inspire new thinking and provide practical pathways for transforming our relationship with nature to ensure humanity's sustainable and just future.

I

Introduction

The concept of Environmentalism—Pathways to Life for Humanity serves as a foundational theme, emphasizing the diverse routes through which individuals and societies can achieve fulfillment, resilience, and sustainability. This section examines the various dimensions of human life, including social, economic, environmental, and spiritual, to present a holistic understanding of what it means to lead a meaningful life. It highlights the interdependence of these dimensions and underscores the importance of a balanced approach to personal and collective well-being (Ekardt, 2019; Gwiszcz, 2018; Bhandari, 2023, 2024).

Pathways to life for humanity concerning sustainability are intricate and multifaceted, encompassing ecological balance, economic resilience, and social equity. At the heart of this concept is recognizing that human life and the natural environment are deeply interconnected. Sustainable pathways ensure that the resources we rely on—clean air, water, fertile soil, and biodiversity—are preserved and regenerated for future generations. This involves adopting practices that minimize environmental degradation, promote renewable energy sources, and enhance the natural ecosystem's capacity to support life.

Economically, sustainable pathways demand a shift from traditional growth models that often exploit natural resources and exacerbate inequalities towards models that value long-term health and resilience. This includes promoting green technologies, supporting circular economies where waste is minimized and resources are reused, and ensuring fair trade practices. Economic sustainability also involves fostering innovation that drives efficiency and reduces ecological footprints, creating jobs that contribute to environmental goals, and investing in infrastructure that can withstand and adapt to climate changes.

Socially, pathways to life for humanity emphasize inclusivity and equity. This means ensuring that all communities have access to the benefits of sustainable practices, such as clean energy, nutritious food, and education on sustainability. It also involves addressing social determinants of health and well-being, such as poverty, inequality, and access to healthcare. Sustainable development goals (SDGs) set by the United Nations serve as a global framework for addressing these issues, aiming to eradicate extreme poverty, reduce inequality, and promote peace and justice while safeguarding the planet.

Cultural and behavioral changes are also critical in the journey toward sustainability. This includes fostering a global consciousness that values sustainability, encouraging lifestyles that reduce waste and carbon footprints, and supporting policies that incentivize sustainable practices. Education and awareness campaigns empower individuals and communities to make informed choices that align with sustainable living.

In essence, pathways to life are about creating systems that balance human needs with the planet's health. This holistic approach requires coordinated efforts across all sectors of society, including government, industry, and civil society, to build a future where humanity can thrive in harmony with the natural world (Ekardt, 2019; Gwiszcz, 2018; Bhandari, 2023, 2024).

I.1 The Pathways of Environmentalism Transformation Toward Ensuring a Sustainable Future for Humanity

The pathways of environmentalism transformation toward ensuring a sustainable future for humanity are multifaceted processes that involve rethinking, restructuring, and reimagining our relationship with the natural world. These pathways are grounded in an integrative and holistic approach to sustainability, addressing environmental preservation, social equity, economic transformation, and ethical responsibility. Environmentalism transformation refers to the profound shifts in mindset, policies, practices, and behaviors necessary to balance human needs with the planet's well-being and ecosystems.

One of the key pathways to environmentalism transformation is the adoption of sustainable development principles, which aim to meet present needs without compromising the ability of future generations to meet their own. This involves prioritizing balanced economic growth within ecological limits and ensuring natural resources are used efficiently and equitably. Sustainable development also emphasizes the integration of social equity, meaning that vulnerable and marginalized communities are not disproportionately affected by environmental degradation or denied access to natural resources. This is essential for fostering global stability and fairness in the fight against environmental crises.

The transformation of environmentalism also includes climate change mitigation and adaptation as a critical pathway. Climate change presents one of the greatest existential risks to humanity and the planet. Addressing this challenge requires reducing greenhouse gas emissions, transitioning to renewable energy sources, and implementing innovative solutions for carbon sequestration. At the same time, adaptation strategies are necessary to build resilience in communities already feeling climate change's impacts. Environmentalism transformation ensures that human societies can thrive in a changing climate while safeguarding ecosystems by focusing on mitigation and adaptation.

Another significant pathway involves biodiversity conservation. Healthy ecosystems are essential for maintaining the planet's life-support systems, from regulating the climate to providing clean water and fertile soil. Nevertheless, human activities such as deforestation, pollution, and habitat destruction push many species to extinction. The pathway of environmentalism transformation promotes the conservation and restoration of biodiversity as a foundation for ecological resilience and sustainability. This includes protecting critical habitats, curbing species overexploitation, and addressing the underlying drivers of biodiversity loss.

Social equity and justice are central to the pathways of environmentalism transformation, as environmental challenges disproportionately impact marginalized communities. Environmental justice seeks to address these disparities by ensuring fair treatment and meaningful involvement of all people in environmental decision-making regardless of race, income, or nationality. This pathway advocates for policies that uplift the voices of underrepresented communities and ensure that the benefits of environmental conservation and sustainability are shared equitably.

Another essential pathway is the transition to a circular economy, which moves away from the traditional linear "take, make, waste" model of resource use. A circular economy emphasizes reducing waste, reusing materials, and recycling products to create a regenerative system where the environmental impact is minimized. This approach helps conserve resources, reduces pollution, lowers greenhouse gas emissions, and fosters sustainable economic growth.

Technological innovation transforms environmentalism by offering solutions to some of the most pressing environmental challenges. From renewable energy technologies to smart agriculture and precision conservation tools, innovation can drive efficiency and minimize human activities' ecological footprint. However, this pathway must be navigated carefully, considering technological advancements' ethical and social implications.

Global cooperation is another critical pathway, as environmental challenges such as climate change, biodiversity loss, and pollution are inherently transboundary issues that require collective action. International agreements, such as the Paris Agreement, and collaborative efforts among governments, civil society, and the private sector are essential for achieving global sustainability goals. This pathway also involves knowledge-sharing and capacity-building across nations, ensuring that all countries, especially the Global South, are equipped to contribute to and benefit from global sustainability efforts.

At the heart of environmentalism transformation is the active engagement of communities in environmental stewardship. Community engagement empowers individuals and local groups to participate in conservation efforts and advocate for sustainable practices within their regions. Environmentalism transformation pathways can create grassroots movements that contribute to the global sustainability agenda by fostering a sense of collective responsibility.

The final pathway in environmentalism transformation involves a comprehensive re-evaluation of nature's role in supporting human well-being through ecosystem services. Recognizing and preserving ecosystems' vital services—such as water purification, air quality regulation, carbon storage, and pollination—is essential for human survival. A transformative approach to environmentalism seeks not only to halt the degradation of ecosystems but also to restore those that have been damaged, ensuring that they can continue to support life for generations to come.

The pathways of environmentalism transformation toward sustainability represent a multidimensional and interconnected framework. These pathways include sustainable development, climate action, biodiversity conservation, social justice, technological innovation, circular economy practices, global cooperation, community engagement, and the preservation of ecosystem services. Together, they form a comprehensive approach to fostering a more just, resilient, and sustainable world in which humanity and nature coexist harmoniously to benefit all life on Earth.

I.2 Why Sustainable Development Is More Critical Than Ever

The importance of sustainable development has never been more apparent. With the planet's resources depleting at an alarming rate and climate-related disasters increasing in frequency and intensity, the stakes have risen for current and future generations. The shift towards renewable energy, responsible consumption, and resource efficiency is crucial to achieving balance, ensuring that economic growth does not come at the expense of environmental degradation or social inequities.

Moreover, sustainability is fundamental for achieving global social equity and justice. Marginalized communities, particularly in developing nations, often bear the brunt of environmental crises, exacerbating poverty and inequality. A sustainable development model prioritizes inclusivity and equity will create a more just world and empower these communities to thrive. The COVID-19 pandemic further highlighted the interconnected nature of global systems, revealing how disruptions in one area can ripple globally. Sustainable development offers a pathway to resilience in an increasingly volatile world.

I.2.1 What we need: Theoretical and practical innovations

A deeper understanding of sustainability requires both theoretical advancements and practical applications. In theory, the field must evolve to incorporate diverse knowledge systems, from Indigenous wisdom to cutting-edge scientific research. Theories of sustainability need to move beyond linear models of progress and consider the complexities of systems thinking ethics and human-nature relationships. Interdisciplinary research integrating environmental science, economics, sociology, and ethics is essential for developing holistic frameworks.

On the practical side, more research is needed to implement sustainable solutions in real-world contexts. This includes exploring how local communities can contribute to global sustainability efforts, how businesses can innovate to align with sustainable development goals, and how governments can create policies that incentivize ecological protection and economic prosperity. Practical innovations, such as regenerative agriculture, circular economy models, and green technologies, will drive sustainability in the coming decades.

I.2.2 How research will bridge the gap

Research is pivotal in shaping sustainable development's theoretical foundations and practical applications. We need more research that identifies new pathways and critically examines the effectiveness of existing strategies. This means fostering collaborative, cross-disciplinary research efforts integrating voices from academia, industry, governments, and local communities.

Additionally, we must focus on data-driven approaches, using advanced technologies like AI and big data to track and analyze environmental and social impacts more effectively. These tools can help us understand complex patterns and relationships within ecosystems and human societies, enabling the creation of more targeted, impactful solutions.

I.2.3 Why we must prioritize research in both theory and practice

Prioritizing research in both theory and practice is vital to achieving sustainable development goals. Without a strong theoretical framework, practical efforts risk being short-sighted, addressing immediate problems without considering long-term implications. Conversely, practical application is necessary for theoretical concepts to remain abstract and connected to the realities on the ground. A balanced approach is needed to ensure that sustainability is not just an academic exercise but a lived experience that benefits humanity.

The future of sustainable development hinges on our ability to innovate both in theory and practice. As we face unprecedented environmental challenges, the need for new approaches to sustainability has never been more pressing. We can create a roadmap for a more just, equitable, and sustainable world for future generations by advancing research in this field.

I.3 Addressing the Growing Demand for Knowledge in Environmentalism

The book is designed to meet the increasing global demand for knowledge in environmentalism and sustainability. As the world grapples with complex environmental challenges like climate change, biodiversity loss, and resource depletion, there is a growing recognition that traditional approaches to environmental management are no longer sufficient. The need for a more comprehensive understanding of these issues, combined with actionable solutions, has never been more urgent.

This book responds directly to this demand by offering a multidimensional approach to environmentalism, integrating ecological, social, economic, and ethical perspectives. It is structured to provide readers with theoretical foundations and practical tools that can be applied in diverse contexts, from academic research and policy development to grassroots activism and corporate sustainability efforts. The knowledge provided here is about understanding the issues and how to take meaningful action toward creating a sustainable future.

I.3.1 Bridging knowledge gaps in sustainability and environmentalism

One of the core objectives of *"Environmentalism: Pathways to Life for Humanity and Sustainability"* is to bridge existing knowledge gaps in sustainability. Environmental challenges are interconnected, yet much of the current research and policy development still needs to be more cohesive, focusing on isolated issues like climate change, deforestation, or pollution. This book presents a holistic perspective, synthesizing these strands into a coherent framework highlighting the interdependence of ecological systems, human societies, and economic activities.

By focusing on interdisciplinary approaches, the book fills a critical gap in the literature where environmentalism intersects with other fields, such as economics, ethics, social justice, and technology. For instance, it explores how ecological economics can offer alternative solutions to mainstream economic models prioritizing profit over planetary health. Similarly, the book addresses the ethical dimensions of sustainability, including intergenerational justice, equity, and humans' moral responsibilities toward the planet and each other.

I.3.2 Equipping readers with actionable insights

Beyond theory, the book is designed to equip readers with practical insights that can be translated into action. In an era where the environmental crisis demands urgent responses, knowledge must be transformative, moving from awareness to implementation. This book offers case studies, examples of successful environmental policies, and innovative practices from around the world that demonstrate how individuals, communities, and organizations can drive change.

This book guides policymakers in developing more effective environmental policies that balance economic growth with ecological preservation. It provides educators and academics with a wealth of resources to shape the next generation of environmental leaders. At the same time, activists and sustainability professionals offer strategies for mobilizing communities and industries toward more sustainable practices. By addressing a broad spectrum of readers, the book ensures that its knowledge can be applied across different sectors, fostering collaboration and collective action.

I.3.3 A comprehensive framework for global sustainability challenges

The book takes a comprehensive approach to addressing the global sustainability challenges that define our times. It explores various topics, including climate change mitigation and adaptation, biodiversity conservation, social equity, renewable energy, and sustainable development, offering readers an integrated framework for understanding these complex issues. This multidimensional perspective is essential for anyone looking to grasp the full scope of the environmental crisis and develop strategies to address its root causes.

The book delves into the relationship between environmental degradation and social inequality, emphasizing that sustainability is not just an environmental issue but a deeply social one. It highlights how marginalized communities, particularly in developing nations, are disproportionately affected by environmental degradation and offers insights into how inclusive environmental policies can promote social justice and equity.

I.3.4 Fostering lifelong learning and transformative education

At the heart of the book is the belief that education is the key to fostering a sustainable future. The book is not just a repository of information but a tool for transformative learning, encouraging readers to rethink their relationship with the natural world and their role in shaping sustainable societies. It promotes a lifelong learning approach, urging individuals, institutions, and communities to adapt continuously to emerging challenges and innovations in sustainability.

The book supports this vision by including diverse ways of knowing, from indigenous knowledge systems to contemporary scientific research, offering a pluralistic view of environmentalism. It encourages readers to engage with these different perspectives and develop a more nuanced understanding of how sustainability can be achieved culturally and contextually appropriately.

I.3.5 A timely contribution to global environmental discourse

This book offers a timely and relevant contribution to the field at a time when environmental issues are at the forefront of global discourse. The world is facing unprecedented challenges, from climate change to loss of biodiversity, and the need for informed, innovative solutions has never been greater. This book is a critical resource for anyone seeking to contribute to the global sustainability movement, providing the theoretical foundation and practical tools needed to make a difference.

The upcoming United Nations Scientist Meeting and global environmental platforms like the IPCC and IPBES will benefit from the insights offered in this book, as they align with the current scientific and policy-oriented discussions on how to transition toward more sustainable societies.

This book positions itself as a thought leader by participating in these global conversations, offering actionable pathways for individuals and institutions.

The book meets the growing demand for knowledge in environmentalism by providing an integrated, interdisciplinary approach to sustainability challenges. It equips readers with the insights and tools they need to create meaningful change, making it an essential resource for academics, policymakers, activists, and anyone committed to building a more sustainable world.

I.4 Theoretical and Practical Problems of the Sustainability Concept

Though widely embraced as essential for long-term environmental and societal well-being, the concept of sustainability faces several theoretical and practical challenges. These problems arise from the complexity of defining sustainability, the difficulty of implementing sustainable practices, and the often conflicting goals of economic growth, environmental protection, and social equity. Understanding and addressing these issues is critical for advancing the sustainability agenda and ensuring its principles are effectively applied in the real world.

I.4.1 Theoretical problems

One of the main theoretical challenges of sustainability is its vague and contested definition. Sustainability is generally understood as meeting the needs of the present without compromising the ability of future generations to meet their own needs. However, this broad definition leaves much room for interpretation. Different fields, stakeholders, and cultures often emphasize different aspects of sustainability—environmental conservation, social equity, or economic viability—leading to competing priorities and disagreements on what constitutes sustainable development. For example, economists may focus on the sustainability of economic growth, while ecologists emphasize the need to maintain biodiversity and natural resources. This lack of a unified definition creates confusion in policy-making and complicates the development of cohesive sustainability strategies.

Another theoretical area for improvement is the difficulty in measuring sustainability. While environmental and economic indicators such as carbon emissions, GDP, or energy consumption can be quantified, social and ethical dimensions of sustainability, like fairness, justice, and well-being, are more complex to assess. This makes it challenging to create metrics that holistically capture sustainability's complex and interconnected nature. Furthermore, because sustainability spans multiple environmental, economic, and social domains, it is often unclear how to balance trade-offs between these areas. For example, an economic policy that boosts growth might lead to environmental degradation, while an environmental conservation effort could negatively impact local employment.

Additionally, the time scale of sustainability presents another theoretical problem. Sustainability is inherently a long-term concept concerned with future generations and the persistence of resources over time. However, decision-makers and stakeholders often operate with short-term goals and incentives, whether in politics, business, or everyday life. This temporal disconnect makes it difficult to create policies and practices that are immediately actionable and aligned with long-term sustainability goals.

I.4.2 Practical problems

In practice, implementing sustainability poses significant challenges due to the complexity of global systems and the diverse interests of stakeholders. One of the most prominent practical issues is the conflict between economic growth and environmental protection. Many industries and governments prioritize short-term economic growth, often achieved through practices that deplete natural resources, pollute ecosystems, and contribute to climate change. Transitioning to sustainable practices usually requires significant investments in renewable energy, waste reduction, and sustainable infrastructure, which can be costly in the short term. This leads to resistance from businesses and governments, particularly in regions where economic development is urgently needed to reduce poverty or unemployment.

Moreover, the unequal distribution of resources exacerbates the challenges of sustainability. Wealthier nations and communities often have the resources to invest in sustainable technologies and infrastructure, while poorer regions face more immediate challenges, such as food security and basic infrastructure. This inequality makes it difficult to implement global sustainability initiatives uniformly. Without addressing these disparities, sustainable practices may deepen existing inequalities by disproportionately benefiting those with access to resources and technology. This raises ethical questions about how sustainability initiatives can be inclusive and equitable across different regions and socioeconomic groups.

Political will and governance are also practical barriers to sustainability. Implementing sustainability policies often requires coordinated efforts across multiple sectors, including governments, corporations, and civil society. However, political systems prioritize immediate concerns, such as economic growth or national security, over long-term environmental sustainability. In addition, differing political ideologies can lead to inconsistent sustainability policies across different administrations or countries. This lack of continuity undermines efforts to create lasting change. Furthermore, global challenges such as climate change require international cooperation, often hampered by conflicting national interests and a lack of global governance mechanisms that can enforce sustainability agreements.

Another practical issue is the need for more current technologies and infrastructures to support sustainable living. While there has been significant progress in areas like renewable energy, many parts of the world still need to rely heavily on fossil fuels and outdated technologies, contributing to environmental degradation. Transitioning to more sustainable technologies requires both financial investment and technical innovation. Still, change is often slow due to institutional inertia, economic constraints, and a lack of awareness or political will.

Lastly, there is the issue of public engagement and behavioral change. Achieving sustainability is not only about changing policies and technologies but also about transforming societal behaviors and values. However, fostering widespread behavioral change is difficult, especially when individuals and communities are not directly affected by the environmental consequences of their actions. Public apathy, lack of awareness, and resistance to change all pose practical barriers to adopting more sustainable lifestyles. Education, communication, and incentives for sustainable behavior will be essential to overcome these obstacles.

The theoretical and practical problems of the sustainability concept highlight the complexities of balancing environmental health, economic growth, and social equity. Theoretical issues like definitional ambiguity, measurement difficulties, and long-term focus must be addressed through more explicit frameworks and interdisciplinary collaboration. Meanwhile, practical challenges such as resource inequality, governance barriers, and technological and behavioral shifts require global cooperation, innovation, and strong political and societal will. Solving these problems is crucial if sustainability becomes a reality and not just an ideal, ensuring a balanced and just future for humanity and the planet.

i.5 Sustainable Development Goals: Why Achieving Them Seems Impossible

The Sustainable Development Goals (SDGs), set by the United Nations in 2015, are a set of 17 ambitious global objectives aimed at addressing a broad spectrum of humanitarian, social, and environmental challenges by 2030. These goals range from eradicating poverty and hunger to ensuring clean energy, reducing inequalities, and addressing climate change. While the vision behind the SDGs is commendable, achieving them seems nearly impossible for several reasons rooted in both practical and systemic hurdles.

I.5.1 Major practical issues and hurdles

I.5.1.1 Resource inequality

One of the most pressing barriers to achieving the SDGs is the deep global inequality in resources, both financial and technological. Many developing countries need help with basic infrastructure issues, from inadequate healthcare systems to poor educational resources and unreliable access to clean water and energy. These countries lack the capital and technology to invest in long-term sustainable development, especially while addressing more immediate needs like poverty reduction and economic stability. Even with international aid, the disparity between resource-rich and resource-poor regions makes implementing uniform global strategies for achieving the SDGs difficult.

I.5.1.2 Political will and governance

There needs to be more political commitment. Achieving the SDGs requires consistent and sustained action across all levels of government, yet many countries prioritize short-term political and economic gains over long-term sustainability goals. Moreover, differing political ideologies lead to inconsistent support for environmental and social policies. Development is often stalled in countries with authoritarian regimes or unstable governments, and resources are diverted toward military spending or corruption rather than invested in SDG-related initiatives. The lack of coordinated global governance mechanisms further weakens efforts to achieve the goals as nations pursue their agendas rather than aligning with global objectives.

I.5.1.3 Economic systems and growth models

The SDGs aim to balance economic growth with environmental sustainability, yet current economic systems are primarily designed for short-term growth, often at the expense of long-term sustainability. The emphasis on GDP growth as a measure of success incentivizes overconsumption, resource extraction, and environmental degradation. This directly conflicts with several SDGs, such as those focused on climate action, life on land, and life below water. Moreover, the global economic system is highly interconnected, meaning that unsustainable practices in one region can have ripple effects on others, making achieving sustainable outcomes in isolation challenging.

I.5.1.4 Climate change and environmental degradation

Climate change is a cross-cutting issue that undermines progress toward nearly all the SDGs. Extreme weather events, rising sea levels, and shifting ecosystems are already exacerbating food insecurity, displacing populations, and putting immense pressure on natural resources. For example, goals related to poverty reduction and food security are directly impacted by environmental changes, as agriculture becomes less predictable and communities reliant on natural resources struggle to adapt. With immediate and coordinated global action to mitigate climate change, the SDGs will stay within reach.

I.5.1.5 Lack of financial commitment

Achieving the SDGs requires an estimated trillions of annual investments, yet financing remains a major hurdle. While some progress has been made in mobilizing resources, public and private investment in the SDGs needs to catch up to what is necessary. Developing countries, in particular, face significant challenges in attracting the investments required for sustainable infrastructure, education, and healthcare. Moreover, global aid often needs to be more allocated, with funding priorities determined by donor interests rather than local needs. This misalignment results in inefficiencies and an inability to meet targets, particularly in the most vulnerable regions.

I.5.1.6 Technological and data gaps

Technology has the potential to accelerate progress toward the SDGs, but many countries need access to the necessary tools and innovations. Technological gaps, particularly in developing nations, hinder advancements in renewable energy, healthcare, education, and infrastructure. Additionally, data gaps make it difficult to track progress toward the SDGs accurately. In many countries, reliable data on issues like poverty, environmental degradation, and health outcomes are either unavailable or inaccurate, leading to poor decision-making and ineffective policy implementation.

1

Reimagining Sustainable Development: Innovative Pathways for an Evolving World

Sustainable development is a holistic approach that aims to meet the needs of the present without compromising the ability of future generations to meet their own needs. This concept, popularized by the 1987 Brundtland Report, encompasses three main pillars: economic growth, environmental protection, and social inclusion. These pillars are interdependent and mutually reinforcing, emphasizing that true progress requires a balanced approach that integrates economic, environmental, and social goals (Bhandari, 2019, 2020, 2024 a,2024b, 2024c; Bhandari and Hanna, 2022; Bhandari and Bhattarai, 2017; Bhandari, P. 2024c; Bhandari and Adhikari, 2024).

Economically, sustainable development seeks to foster growth that is both inclusive and resilient. This involves creating opportunities for decent work, fostering innovation, and ensuring that economic activities do not lead to environmental degradation. Sustainable economic practices include investing in green technologies, promoting energy efficiency, and supporting industries that reduce waste and utilize resources more effectively. It also involves creating economic systems that are fair and equitable, ensuring that the benefits of growth are distributed broadly across society and that vulnerable populations are not left behind.

Environmental protection is a critical component of sustainable development. This entails safeguarding natural resources, reducing pollution, and mitigating climate change. Strategies for environmental sustainability include transitioning to renewable energy sources, conserving biodiversity, and promoting sustainable agriculture and forestry practices. It also involves protecting ecosystems that provide essential services, such as clean water, air, and fertile soil, vital for human health and well-being. Addressing environmental challenges requires international cooperation and implementing policies that incentivize sustainable practices and penalize harmful activities.

Social inclusion is the third pillar of sustainable development, emphasizing the need to create just, equitable, and inclusive societies. This involves ensuring access to essential services such as education, healthcare, and clean water, promoting gender equality, and reducing inequalities within and between countries. Social sustainability also includes protecting human rights, fostering participatory governance, and building resilient communities that can adapt to changing conditions. Empowering marginalized groups and ensuring all voices are heard in decision-making is key to achieving social sustainability.

The United Nations' Sustainable Development Goals (SDGs) provide a comprehensive global framework for advancing sustainable development. The 17 goals address many issues, from poverty and hunger to climate action and peace, reflecting the interconnected nature of sustainable development. Achieving these goals requires coordinated efforts from governments, businesses, civil society, and individuals and robust data and monitoring systems to track progress and hold stakeholders accountable.

Sustainable development creates a future where economic prosperity, environmental health, and social well-being are mutually reinforcing and balanced. It requires a long-term perspective and a commitment to making decisions that benefit current and future generations (Bhandari, 2019, 2020, 2024a, 2024b, 2024c; Bhandari and Hanna, 2022; Bhandari and Bhattarai, 2017; Bhandari, P. 2024c; Bhandari and Adhikari, 2024). By embracing sustainable development, societies can work towards a world where everyone can thrive within the planet's ecological limits.

1.1 Balanced Growth

Promote economic growth that meets present needs without compromising the ability of future generations to meet theirs. Balanced growth is an essential principle of sustainable development, focused on promoting economic expansion that satisfies current needs without jeopardizing the prospects of future generations. This concept recognizes that economic growth, while vital for improving living standards and reducing poverty, must be managed so as not to deplete natural resources, degrade the environment, or exacerbate social inequalities. Achieving balanced growth requires integrating economic, environmental, and social considerations into development strategies and policies (Mennillo et al., 2011; Behdadfar and Samaei, 2024; Edwards, 2021).

Economically, balanced growth involves fostering industries and practices that are both productive and sustainable. This includes investing in green technologies and renewable energy sources that reduce reliance on fossil fuels and minimize environmental impact. It also means promoting efficient resource use and waste reduction through circular economy models, where products are designed for longevity, reuse, and recycling. By prioritizing sustainable production and consumption patterns, economies can grow while preserving the natural capital essential for future prosperity.

Environmental sustainability is a cornerstone of balanced growth. Economic activities must be aligned with ecological constraints to ensure that natural resources, such as water, air, and soil, remain available and healthy for future generations. This involves implementing stringent environmental regulations, protecting biodiversity, and restoring ecosystems that human activities have degraded. Policies incentivizing businesses to adopt sustainable practices, such as reducing emissions and conserving resources, are crucial for maintaining the ecological balance necessary for long-term economic stability.

Social equity is another critical aspect of balanced growth. Ensuring that economic benefits are broadly shared and no one is left behind is essential for creating resilient and inclusive societies. This requires addressing structural inequalities, providing access to quality education and healthcare, and promoting decent work opportunities. Social safety nets and policies supporting vulnerable populations are also vital for ensuring everyone can participate in and benefit from economic growth. Balanced growth contributes to social cohesion and stability by fostering social inclusion and reducing disparities, which are fundamental for sustained development.

Innovation and technological advancement play a crucial role in achieving balanced growth. Research and development in sustainable technologies can drive economic growth while minimizing environmental impact. This includes advancements in energy efficiency, sustainable agriculture, and green infrastructure. Governments and businesses must invest in innovation and create an enabling environment for adopting new technologies that support sustainable development goals. Collaboration between public and private sectors and international cooperation is essential for scaling up successful innovations and spreading their benefits globally.

Balanced growth creates synergy between economic development, environmental stewardship, and social well-being. It requires a comprehensive approach integrating sustainability into all economic planning and policy-making aspects (Mennillo et al., 2011; Behdadfar and Samaei, 2024; Edwards, 2021). By prioritizing long-term sustainability over short-term gains, societies can build robust economies that support current needs while ensuring that future generations inherit a world capable of sustaining their growth and well-being. This holistic approach to development is essential for achieving a sustainable and equitable future for all.

1.2 Resource Efficiency

Resource efficiency is a cornerstone of sustainable development. It focuses on optimizing the use of natural resources to minimize waste and reduce environmental impact. This concept is essential for creating sustainable systems that balance economic growth with ecological integrity. By using resources more efficiently, societies can achieve more with less, reducing the strain on the environment and ensuring that natural resources remain available for future generations.

At its core, resource efficiency involves maximizing the productivity of resources throughout their lifecycle—from extraction and production to consumption and disposal. This can be achieved through various strategies, such as improving manufacturing processes to use less material and energy, designing products for durability and recyclability, and promoting practices that extend the lifespan of goods. For instance, adopting circular economy principles, where products and materials are reused, refurbished, and recycled, can significantly enhance resource efficiency and reduce waste (Lee, 2015; Zakari et al., 2022; Gerstlberger et al., 2014).

Resource efficiency means adopting cleaner production techniques and technologies that reduce resource input and waste output in industry and manufacturing. This includes using advanced technologies that improve energy efficiency, reduce water usage, and minimize emissions. Industries can also implement waste-to-resource practices, where by-products and waste materials are repurposed or recycled into new products. Such practices conserve resources, reduce costs, and improve competitiveness by turning waste into valuable inputs.

The transition to renewable energy sources is a critical aspect of resource efficiency. Renewable energy, such as solar, wind, and hydroelectric power, offers a sustainable alternative to fossil fuels, reducing greenhouse gas emissions and conserving finite natural resources. By investing in renewable energy infrastructure and technologies, societies can decrease their reliance on non-renewable resources and lower their environmental footprint. Energy efficiency measures, such as improving building insulation, upgrading to energy-efficient appliances, and optimizing industrial energy use, enhance resource efficiency by reducing overall energy demand.

Resource efficiency also extends to water and land use. Sustainable water management practices, such as efficient irrigation techniques, wastewater recycling, and reducing water wastage, are crucial for conserving this vital resource. Precision farming, agroecology, and organic farming can improve resource use efficiency, enhance soil health, and reduce chemical inputs, leading to more sustainable food production systems. Urban planning incorporating green infrastructure, such as parks, green roofs, and permeable surfaces, can optimize land use, enhance biodiversity, and improve the quality of life in cities.

Consumers play a vital role in promoting resource efficiency. By making informed choices and adopting sustainable consumption habits, individuals can drive demand for resource-efficient products and services. This includes supporting companies with sustainable practices, reducing single-use plastics, recycling, and choosing products with minimal packaging (Lee, 2015; Zakari et al., 2022; Gerstlberger et al., 2014). Public awareness campaigns and education on sustainability can empower consumers to make decisions that align with resource efficiency goals.

Optimizing the use of natural resources to minimize waste and reduce environmental impact is essential for sustainable development. Resource efficiency conserves finite resources, protects the environment, and drives economic benefits by reducing costs and fostering innovation. Achieving resource efficiency requires collaboration across industries, governments, and consumers, supported by policies and practices prioritizing sustainability. By embracing resource efficiency, societies can build resilient, sustainable systems that support both present and future generations.

1.3 Renewable Energy

To reduce greenhouse gas emissions, we must transition from fossil fuels to renewable energy sources like solar, wind, and hydropower. The transition from fossil fuels to renewable energy sources represents a crucial strategy for combating climate change and reducing greenhouse gas emissions. This shift is pivotal for creating a sustainable energy future and mitigating the environmental impacts associated with the extraction and burning of fossil fuels such as coal, oil, and natural gas. Renewable energy sources like solar, wind, and hydropower offer cleaner alternatives that significantly reduce our carbon footprint and enhance environmental stewardship (Lund, 2007; Østergaard et al., 2022; Güney, 2019).

Solar energy harnesses the sun's power through photovoltaic cells or solar thermal systems. It is one of the most abundant and widely available renewable energy sources. The technology has seen significant efficiency and cost reduction advancements over the past decade, making solar power more accessible and economically viable. Solar panels can be installed on residential rooftops, commercial buildings, and large-scale solar farms, providing a decentralized and flexible energy solution. By capturing and converting sunlight into electricity, solar energy reduces reliance on fossil fuels and decreases greenhouse gas emissions.

Wind energy is another key player in the renewable energy landscape. Wind turbines convert the kinetic energy of wind into electrical power. Wind farms, located onshore or offshore, can generate substantial amounts of electricity and have become a significant component of many countries' renewable energy strategies. Wind energy is clean, abundant, and can be deployed at various scales, from small residential turbines to large utility-scale projects. Advancements in turbine technology and increased efficiency have made wind power a cost-effective alternative to fossil fuels, contributing to significant reductions in carbon emissions.

Hydropower, or hydroelectric power, generates electricity by harnessing the energy of flowing water. This can be achieved through large-scale dams, small-scale run-of-river projects, or pumped storage systems. Hydropower has been a longstanding renewable energy source and provides a reliable and consistent electricity supply. It also offers additional benefits, such as water storage for irrigation and flood control. However, it is essential to manage hydropower projects carefully to minimize ecological impacts on aquatic ecosystems and local communities.

The transition to renewable energy also involves integrating these sources into existing energy systems and infrastructure. This includes enhancing grid stability and capacity to accommodate variable energy inputs from solar and wind sources, developing energy storage solutions to balance supply and demand, and investing in smart grid technologies. Energy storage, such as batteries and pumped hydro storage, plays a critical role in ensuring a consistent and reliable supply of renewable energy by storing excess power generated during peak production times and releasing it during periods of high demand.

Policy support and investment are crucial for accelerating the transition to renewable energy. Governments can play a significant role by implementing supportive policies, such as tax incentives, subsidies for renewable energy projects, and regulations that limit carbon emissions. Public and private sector investments in research and development can drive innovation, reduce costs, and increase the adoption of renewable technologies. Additionally, international cooperation and agreements, such as the Paris Agreement, help align global efforts toward a low-carbon future and encourage sharing best practices and technologies.

Public awareness and engagement are also crucial for the transition to renewable energy. Educating individuals about the benefits of renewable energy, promoting energy efficiency, and encouraging sustainable practices can drive consumer demand and support for clean energy solutions. Community-based renewable energy projects and initiatives can empower local populations to participate in and benefit from the energy transition, fostering greater acceptance and commitment to sustainable energy practices.

Transitioning from fossil fuels to renewable energy sources is essential for reducing greenhouse gas emissions and addressing climate change. Solar, wind, and hydropower offer viable and sustainable alternatives that significantly lower our carbon footprint and promote environmental conservation (Lund, 2007; Østergaard et al., 2022; Güney, 2019). Achieving this transition requires coordinated efforts across technology development, policy support, infrastructure investment, and public engagement to build a resilient and sustainable energy future.

2

Climate Change Mitigation

Climate change mitigation refers to the array of strategies and actions designed to reduce or prevent the emission of greenhouse gases (GHGs), thereby slowing global warming and mitigating the adverse effects of climate change. Central to this effort is the transition from fossil fuels to renewable energy sources such as solar, wind, and hydropower. These energy sources produce little to no GHG emissions compared to coal, oil, and natural gas, making them crucial in the fight against climate change (Raman et al., 2012; Fawzy et al., 2020; Abanades et al., 2017). By investing in clean energy technologies and infrastructure, societies can significantly cut emissions while fostering economic growth and creating jobs.

Energy efficiency is another vital component of climate change mitigation. Improving energy use efficiency across various sectors—such as residential buildings, transportation, and industry—can substantially lower energy consumption and emissions. Upgrading to energy-efficient appliances, enhancing building insulation, and adopting advanced industrial processes help reduce the demand for fossil fuels and associated emissions. In transportation, shifting towards electric vehicles, expanding public transit options, and promoting alternative modes of transport like cycling can further decrease GHG emissions from one of the most significant contributors to climate change.

Sustainable practices play a critical role in agriculture and land use. Precision farming, agroforestry, and organic farming reduce land use emissions and enhance natural systems' ability to sequester carbon. Protecting and restoring forests and wetlands, which act as natural carbon sinks, is also essential. Carbon pricing mechanisms, including carbon taxes and cap-and-trade systems, incentivize emission reductions by attaching a cost to carbon emissions, encouraging businesses and individuals to adopt cleaner technologies.

Global cooperation is imperative for effective climate change mitigation. International agreements like the Paris Agreement set targets for reducing global GHG emissions and provide a framework for collective action.

Countries are encouraged to establish and meet their national targets while supporting one another in achieving global climate goals. Public awareness and engagement further drive support for climate policies and promote sustainable practices at the individual and community levels. Climate change mitigation requires a multifaceted approach involving technological innovation, policy support, and widespread public participation to create a sustainable future (Raman et al., 2012; Fawzy et al., 2020; Abanades et al., 2017).

2.1 Emission Reductions

Implement policies and technologies to reduce carbon emissions across industries. Emission reductions are pivotal in tackling climate change and involve implementing a combination of policies and technologies to decrease carbon emissions across various sectors. As the primary driver of global warming, carbon emissions need to be curtailed significantly to mitigate adverse environmental impacts and meet international climate goals. A multifaceted approach, integrating regulatory measures, technological advancements, and sector-specific strategies, is essential for achieving these reductions (Daba and Dejene, 2018; Anup et al., 2013; Bassi and Brink, 2013).

Policies play a crucial role in steering industries toward lower carbon emissions. Governments can establish regulations that set industry emission limits, mandate cleaner production processes, and enforce stricter environmental standards. Carbon pricing mechanisms, such as carbon taxes and cap-and-trade systems, create economic incentives for reducing emissions by assigning a cost to carbon emissions. These mechanisms encourage companies to invest in cleaner technologies and adopt practices that lower their carbon footprint. Additionally, subsidies and financial incentives for renewable energy projects and efficiency improvements can further support emission reduction efforts.

Technologies are integral to reducing emissions across industries. Innovations such as carbon capture and storage (CCS) capture CO2 emissions from industrial processes and store them underground, preventing them from entering the atmosphere. Advances in renewable energy technologies, such as more efficient solar panels and wind turbines, provide cleaner alternatives to fossil fuels. Energy efficiency technologies, including smart grids and energy-efficient machinery, help industries reduce their energy consumption and associated emissions. In manufacturing, improvements in processes and materials can lower emissions, while in transportation, electric and hybrid vehicles offer alternatives to traditional fossil fuel-powered transportation.

Sector-specific strategies are also critical for effective emission reductions. In the power sector, transitioning to renewable energy sources and increasing the share of low-carbon power generation is essential. The industrial sector can focus on adopting cleaner technologies and optimizing processes to minimize emissions. The agricultural industry can implement sustainable practices such as precision farming and improved livestock management to reduce methane emissions and enhance carbon sequestration. In the transportation sector, promoting public transit, cycling, and walking, alongside the adoption of electric vehicles, can significantly reduce emissions from this major source.

Achieving meaningful emission reductions ultimately requires a coordinated effort among governments, industries, and individuals. Policies and technologies must work in tandem to drive substantial and sustained reductions in carbon emissions. By fostering innovation, implementing effective regulations, and supporting sector-specific strategies, societies can make significant strides toward mitigating climate change and building a more sustainable future.

2.2 Carbon Sequestration

Enhance natural and technological methods to capture and store atmospheric CO2. Carbon sequestration is a crucial strategy for mitigating climate change. It focuses on capturing and storing atmospheric CO2 to reduce greenhouse gas concentrations and combat global warming. This approach encompasses natural and technological methods, each significantly addressing the challenge of excess carbon dioxide in the atmosphere (Daba and Dejene, 2018; Anup et al., 2013; Bassi and Brink, 2013).

Carbon sequestration enhances the Earth's natural processes that capture and store carbon. Forests, wetlands, grasslands, and oceans act as significant carbon sinks by absorbing CO2 from the atmosphere through photosynthesis and oceanic absorption. Reforestation and afforestation—planting trees on previously deforested or non-forested lands—can increase the capacity of forests to sequester carbon. Protecting existing forests, particularly tropical rainforests, is equally crucial, as these ecosystems store vast amounts of carbon and contribute significantly to global carbon sinks. Restoring wetlands and peatlands, which store large quantities of carbon in their waterlogged soils, can further enhance natural carbon sequestration. Sustainable agricultural practices, such as no-till farming and cover cropping, also contribute by increasing soil organic matter and improving soil carbon storage.

Technological carbon sequestration involves engineered systems designed to capture CO2 emissions directly from industrial sources or the atmosphere. One prominent technology is Carbon Capture and Storage (CCS), which captures CO2 emissions from power plants and industrial facilities, compresses them, and transports them to underground storage sites. These sites, such as depleted oil and gas fields or deep saline aquifers, provide secure long-term storage to prevent CO2 from re-entering the atmosphere. Another technological approach is Direct Air Capture (DAC), which uses chemical processes to capture CO2 directly from the ambient air. Once captured, the CO2 can be stored underground or used in various applications, such as enhanced oil recovery or the production of synthetic fuels.

Integration and scale are crucial to maximizing the effectiveness of carbon sequestration. Combining natural and technological methods can create a more comprehensive approach to reducing atmospheric CO2. For instance, integrating reforestation with CCS technologies can enhance overall carbon removal efforts. Scaling up these methods to meet global climate goals requires substantial investment, research, and development to improve efficiency and reduce costs. Policy support, such as carbon pricing and incentives for sequestration projects, is essential for driving the adoption and expansion of natural and technological methods.

Carbon sequestration—through natural processes and technological innovations—is vital for reducing atmospheric CO2 and mitigating climate change (Daba and Dejene, 2018; Anup et al., 2013; Bassi and Brink, 2013). By enhancing natural carbon sinks, implementing advanced technologies, and integrating these approaches, societies can make significant progress toward stabilizing global temperatures and achieving long-term climate goals.

2.3 Climate Adaptation

Develop infrastructure and practices that help communities adapt to changing climate conditions. Climate adaptation involves developing infrastructure and practices that enable communities to adjust to the impacts of climate change, thereby reducing vulnerability and enhancing resilience. As climate change leads to more frequent and severe weather events, such as floods, heatwaves, and storms, adaptation strategies become essential for protecting lives, properties, and ecosystems. Effective adaptation requires a multifaceted approach integrating planning, infrastructure development, and community engagement (Dessai and Hulme, 2004; Massetti and Mendelsohn, 2018; Shi and Moser, 2021).

Infrastructure development is a cornerstone of climate adaptation. Building resilient infrastructure involves designing and constructing systems that withstand extreme weather events and changing climate conditions. For example, improving stormwater management systems can help prevent flooding by enhancing drainage and incorporating green infrastructure such as permeable pavements and rain gardens. Coastal communities can invest in seawalls, levees, and natural barriers like mangroves and wetlands to protect against sea-level rise and storm surges. Upgrading infrastructure to withstand higher temperatures, such as heat-resistant materials for roads and buildings, can also mitigate the impacts of heat waves.

Urban planning and land use play critical roles in climate adaptation. Implementing zoning regulations and land-use policies that consider climate risks can help reduce exposure to hazards. For instance, avoiding construction in flood-prone areas and incorporating green spaces into urban designs can mitigate the effects of extreme weather. Developing heat island mitigation strategies, such as increasing urban greenery and reflective surfaces, can help lower city temperatures. Promoting sustainable building practices, including energy-efficient designs and renewable energy, enhances resilience and reduces environmental impacts.

Community practices and preparedness are vital for effective adaptation. Educating and empowering communities to understand climate risks and adopt adaptive behaviors can enhance resilience at the local level. This includes developing emergency response plans, conducting risk assessments, and engaging in community-based adaptation projects. Strengthening social networks and support systems can help communities recover more effectively from climate-related disruptions. Encouraging local adaptation initiatives, such as community gardens and disaster preparedness drills, fosters a proactive approach to managing climate risks.

Ecosystem-based adaptation involves leveraging natural systems to provide resilience against climate impacts. Protecting and restoring ecosystems such as wetlands, forests, and coastal mangroves can enhance natural defenses against flooding, erosion, and heat waves. These ecosystems provide crucial services and contribute to carbon sequestration and environmental health. Integrating ecosystem-based approaches with traditional infrastructure solutions can create a more comprehensive and sustainable adaptation strategy.

Policy and governance are essential for supporting climate adaptation efforts. Governments can develop and implement policies that promote resilience and allocate resources for adaptation projects. This includes integrating climate adaptation into planning processes, providing funding and technical support for adaptation initiatives, and fostering collaboration between different sectors and levels of government. International cooperation and knowledge sharing can also help countries learn from each other's experiences and implement effective adaptation strategies.

Climate adaptation involves a proactive approach to managing the impacts of climate change by developing resilient infrastructure, implementing effective land-use planning, fostering community engagement, and leveraging natural systems (Dessai and Hulme, 2004; Massetti and Mendelsohn, 2018; Shi and Moser, 2021). By integrating these strategies and supporting them through policy and governance, societies can enhance their ability to adapt to changing climate conditions and build a more resilient future.

3

Biodiversity Conservation

Biodiversity refers to the variety and variability of life forms on Earth, encompassing species diversity, ecosystems, and genetic variations within species. It includes all living organisms, from the smallest microorganisms to the largest mammals, and the complex interactions they form within their environments. Biodiversity is fundamental to the health and stability of ecosystems, providing essential services such as pollination, nutrient cycling, and climate regulation. It supports the resilience of ecosystems by enabling them to adapt to changes and recover from disturbances. The richness of biodiversity is not just about the number of species but also their genetic diversity and the ecosystems they form, which contribute to the overall functionality and sustainability of the natural world (Miller et al., 2009; Bhandari, 2022, 2024).

Biodiversity conservation protects and maintains the diversity of life on Earth to ensure that ecosystems remain healthy and resilient. Conservation efforts aim to preserve the variety of species, their habitats, and the ecological processes that sustain them. This involves a range of strategies, from establishing protected areas and wildlife reserves to restoring degraded habitats and implementing sustainable management practices. Conservation also focuses on preventing the extinction of endangered species through efforts such as habitat protection, legal regulations, and captive breeding programs (Bhandari, 2022).

One key aspect of biodiversity conservation is protecting habitats. Many species are threatened by habitat loss due to human activities such as deforestation, urbanization, and agriculture. Protecting natural habitats and creating corridors to connect fragmented landscapes are essential for maintaining biodiversity. Restoration projects that rehabilitate degraded environments can also help restore ecological balance and support species recovery.

Biodiversity conservation addresses threats like climate change, pollution, and resource overexploitation. Climate change can alter habitats and disrupt species distributions, while pollution can degrade ecosystems and harm wildlife. Sustainable agriculture, fishing, and forestry practices are crucial for minimizing human impacts on the environment and ensuring that natural resources are used responsibly.

Community involvement and education are also vital for effective biodiversity conservation. Engaging local communities in conservation efforts can foster stewardship and ensure conservation strategies align with local needs and values. Education and awareness programs help people understand the importance of biodiversity and encourage behaviors that support conservation goals.

Biodiversity encompasses the wide variety of life forms and their interactions within ecosystems, and its conservation is crucial for maintaining the health and resilience of our planet. Effective conservation strategies involve protecting habitats, addressing environmental threats, and engaging communities to preserve biodiversity for future generations (Miller et al., 2009; Bhandari, 2022, 2024).

3.1 Habitat Protection

Preserve and restore natural habitats to support diverse ecosystems. Habitat protection is a critical aspect of biodiversity conservation. It focuses on preserving and restoring natural environments to support diverse ecosystems and the myriad species they harbor. By safeguarding habitats, we ensure that ecosystems remain functional and resilient, providing essential services and sustaining a wide range of life forms (Shi et al., 2005; Murray et al., 1997; Vuorisalo and Laihonen, 2000).

Preserving natural habitats involves setting aside and managing protected areas such as national parks, wildlife reserves, and marine protected areas. These areas serve as sanctuaries for species, offering refuge from threats such as habitat destruction, pollution, and overexploitation. Effective management of these protected areas includes enforcing regulations to prevent illegal activities, controlling invasive species, and monitoring ecosystem health. By maintaining large, contiguous areas of habitat, we can support populations of species that require extensive territories or specialized conditions.

Restoring degraded habitats is equally essential for supporting biodiversity. Habitat degradation, caused by deforestation, urbanization, and agriculture, can lead to the loss of essential ecological functions and threaten species' survival. Restoration projects aim to rehabilitate these areas by reintroducing native plant species, removing invasive species, and restoring natural processes like water flow and soil fertility. Successful restoration improves the area's ecological integrity and enhances its capacity to support wildlife and provide ecosystem services.

Habitat connectivity is another critical aspect of habitat protection. Fragmentation of habitats into isolated patches can hinder species movement and reduce genetic diversity. Creating ecological corridors and stepping stones—such as wildlife bridges or greenways—can link fragmented habitats, allowing species to migrate, find resources, and reproduce. These connections are vital for maintaining healthy populations and enabling species to adapt to environmental changes.

Community involvement is essential for adequate habitat protection and restoration. Local communities often rely on natural resources for their livelihoods and can play a crucial role in conservation efforts. Engaging communities in habitat management, restoration projects, and sustainable practices ensures conservation efforts align with local needs and fosters a sense of stewardship. Education and awareness programs can also help communities understand the value of habitats and the benefits of conservation.

Policy support and regulations are crucial for safeguarding habitats. Governments and organizations can implement policies that protect critical habitats, regulate land use, and promote sustainable development. International agreements and conventions, such as the Convention on Biological Diversity, provide frameworks for collaborative habitat protection and conservation efforts (Shi et al., 2005; Murray et al., 1997; Vuorisalo and Laihonen, 2000).

Habitat protection is vital for sustaining diverse ecosystems and the species they support. By preserving natural areas, restoring degraded habitats, enhancing connectivity, involving communities, and implementing supportive policies, we can ensure that ecosystems remain healthy and resilient, providing essential benefits for wildlife and human populations.

3.2 Species Conservation

Implement strategies to protect endangered species and prevent biodiversity loss. Species conservation is a fundamental aspect of biodiversity preservation. It focuses on implementing strategies to protect endangered species and prevent biodiversity loss. As species face threats from habitat destruction, climate change, poaching, and other human activities, targeted conservation efforts are essential for ensuring their survival and maintaining ecosystem health (Frankham et al., 2012; Paoli et al., 2010; Rawat and Agarwal, 2015).

Identifying and prioritizing endangered species is a crucial first step in species conservation. Organizations and governments assess the conservation status of species through tools like the International Union for Conservation of Nature (IUCN) Red List, which categorizes species based on their risk of extinction. By identifying species at greatest risk, conservation efforts can be directed where they are most needed. Prioritization often involves evaluating factors such as population size, habitat range, and specific threats each species faces.

Habitat protection and restoration are central to conserving endangered species. Many species are threatened by habitat loss due to deforestation, urban expansion, and agricultural activities. Protecting critical habitats by establishing reserves and protected areas helps ensure that species have the space and resources they need to survive. Restoration projects, which involve rehabilitating degraded habitats, can also provide essential support for species recovery by improving the quality and connectivity of their environments.

Anti-poaching and wildlife trade regulations are essential for protecting species from illegal activities. Poaching and the illicit wildlife trade pose significant threats to many endangered species, particularly those targeted for their body parts or as pets. Implementing and enforcing stringent laws, increasing surveillance and anti-poaching patrols, and promoting sustainable and legal wildlife trade practices are crucial for curbing these threats. International cooperation and agreements, such as the Convention on International Trade in Endangered Species of Wild Fauna and Flora (CITES), help regulate and monitor wildlife trade.

Captive breeding and reintroduction programs are valuable tools for species conservation. Captive breeding involves breeding endangered species in controlled environments, such as zoos or specialized facilities, to increase their populations and genetic diversity. Reintroduction programs aim to release these individuals into their natural habitats once suitable conditions are suitable. Success in these programs depends on careful planning, including habitat assessment, disease management, and monitoring post-release survival and adaptation.

Community involvement and education play a vital role in species conservation. Engaging local communities in conservation efforts can foster support and create incentives for protecting endangered species. Education programs that raise awareness about the importance of biodiversity and the specific needs of endangered species can encourage conservation-friendly behaviors and practices. Community-based conservation initiatives, such as wildlife-friendly agriculture and eco-tourism, provide alternative livelihoods while supporting conservation goals.

Research and monitoring are essential for effective species conservation. Ongoing research helps to understand species' ecological requirements, population dynamics, and threats' impacts. Monitoring programs track population trends, habitat conditions, and the effectiveness of conservation measures. This information is crucial for adapting and improving conservation strategies to address emerging challenges and ensure the long-term survival of endangered species (Frankham et al., 2012; Paoli et al., 2010; Rawat and Agarwal, 2015).

Species conservation involves a comprehensive approach to protecting endangered species and preventing biodiversity loss. By prioritizing species at risk, protecting and restoring habitats, enforcing anti-poaching measures, utilizing captive breeding programs, engaging communities, and supporting research, we can work towards preserving the rich diversity of life on Earth for future generations.

3.3 Sustainable Land Use

Promote agricultural and forestry practices that maintain ecosystem health. Sustainable land use is essential for maintaining ecosystem health while meeting human food, resources, and livelihood needs. It involves promoting agricultural and forestry practices that protect and enhance the environment, ensure long-term productivity, and support the well-being of communities. Sustainable land use balances development with conservation and resilience by integrating environmental, economic, and social considerations (Young et al., 2005; Borón et al., 2016; Kiss, 2004).

Sustainable agricultural practices focus on managing land to maintain soil fertility, minimize environmental impact, and support biodiversity. Crop rotation and polyculture enhance soil health and reduce pest and disease pressures by diversifying crops. Organic farming methods, which avoid synthetic pesticides and fertilizers, promote using natural inputs and improve soil and water quality. Conservation tillage reduces soil erosion and preserves soil structure by minimizing the disturbance of soil layers. Agroforestry, the practice of integrating trees and shrubs into agricultural systems, provides benefits such as improved soil fertility, increased biodiversity, and enhanced water management.

Sustainable forestry practices aim to manage forest resources in a way that maintains their ecological functions and supports local communities. Selective and reduced-impact logging techniques minimize damage to the surrounding forest and maintain habitat integrity. Certification programs like those offered by the Forest Stewardship Council (FSC) ensure that forestry operations meet environmental and social standards. Reforestation and afforestation efforts help restore degraded lands, enhance carbon sequestration, and rebuild biodiversity. Sustainable forest management protects ecosystems from illegal logging, land conversion, and other threats.

Land use planning is a crucial component of sustainable land use. By integrating environmental considerations into land use decisions, planners can avoid placing agricultural and forestry activities in ecologically sensitive areas, such as wetlands and critical habitats. Zoning regulations can direct development away from areas prone to erosion or flooding, while conservation easements can protect critical natural areas from conversion to other uses. Effective land use planning involves engaging stakeholders, including local communities, to ensure land use decisions reflect their needs and priorities.

Soil and water management are vital for sustainable land use. Practices that conserve soil moisture, such as mulching and contour plowing, help reduce water use and prevent soil erosion. Integrated water management ensures water resources are used efficiently and sustainably, balancing agricultural needs with environmental protection. Techniques such as rainwater harvesting and creating buffer zones around water bodies help maintain water quality and reduce runoff.

Economic incentives and support play a significant role in promoting sustainable land use. Financial mechanisms such as subsidies for sustainable practices, payment for ecosystem services programs, and market-based incentives can encourage land users to adopt environmentally friendly practices. Providing access to technical assistance and education helps farmers and foresters implement sustainable techniques and understand their benefits.

Climate change adaptation is also a critical aspect of sustainable land use. Practices that enhance soil health, increase vegetation cover, and improve water management can help landscapes adapt to changing climate conditions (Young et al., 2005; Borón et al., 2016; Kiss, 2004). By building resilience through sustainable practices, land use systems can better withstand and recover from climate-related stresses.

Sustainable land use involves promoting agricultural and forestry practices that protect and enhance ecosystem health while meeting human needs. We can balance development and conservation by adopting techniques that preserve soil and water quality, engaging in effective land use planning, and providing economic and technical support. This approach ensures land-based systems' long-term productivity and resilience, supporting environmental sustainability and human well-being.

4

Social Equity and Justice

Social equity and justice are foundational principles for creating fair and inclusive societies. They ensure that all individuals have equal opportunities and access to resources regardless of their background. Social equity emphasizes the need to address disparities and support marginalized and disadvantaged groups to achieve fairness in outcomes. Social justice, closely related, focuses on the ethical distribution of resources, opportunities, and privileges, striving to correct systemic inequalities and ensure that every person's rights and dignity are upheld (Hart, 1974; Greenberg and Cohen, 2014; Cahalan et al., 2024; Bhandari and Adhikari, 2024).

Addressing disparities involves recognizing and rectifying the uneven distribution of resources and opportunities that affect marginalized communities. This includes tackling issues such as income inequality, educational disparities, and unequal access to healthcare. Policies and programs designed to support disadvantaged groups—such as targeted educational funding, affordable healthcare initiatives, and social safety nets—play a crucial role in reducing these disparities. By focusing on the needs of those who are most disadvantaged, societies can work towards more equitable outcomes and improve overall social cohesion.

Equitable resource distribution is another key aspect of social equity and justice. It requires ensuring all individuals have fair access to essential resources like housing, clean water, and employment opportunities. This often involves addressing historical injustices and systemic barriers that have led to unequal access. For example, implementing affordable housing policies, investing in underserved communities, and promoting fair labor practices can help address these inequalities (Bhandari and Adhikari, 2024). Moreover, efforts to ensure environmental justice—by addressing disproportionate environmental impacts on marginalized communities—are essential for achieving broader social equity.

Participatory governance is fundamental to social equity and justice. Inclusive decision-making processes involving diverse communities can help ensure that policies and programs reflect the needs and perspectives of all citizens. Encouraging civic engagement and providing platforms for marginalized voices to be heard contribute to more equitable and effective governance. Additionally, promoting transparency and accountability in government and institutional practices helps prevent discrimination and corruption, ensuring that resources and opportunities are distributed fairly.

Social justice also involves addressing and dismantling systemic and institutional biases perpetuating inequality. This includes challenging discriminatory practices and structures within various systems, such as education, criminal justice, and employment. Promoting diversity and inclusion within institutions and creating policies that actively counteract discrimination are crucial to achieving social justice. Furthermore, fostering a culture of respect and mutual understanding helps build more inclusive and harmonious communities (Bhandari and Adhikari, 2024).

Social equity and justice are essential for fostering fair and inclusive societies. By addressing disparities, ensuring equitable resource distribution, promoting participatory governance, and dismantling systemic biases, societies can work towards achieving greater fairness and ensuring that all individuals have the opportunity to thrive. These principles enhance individual well-being and contribute to overall social stability and cohesion.

4.1 Inclusive Policies

Formulate policies that ensure equitable access to resources and opportunities for all societal groups. Inclusive policies are essential for fostering equitable access to resources and opportunities across all societal groups. These policies aim to address and mitigate inequalities, ensuring that marginalized and disadvantaged communities receive the support they need to achieve fair outcomes. By integrating principles of inclusivity and equity into policy formulation, governments and organizations can create environments where everyone can succeed and contribute to society (Shaeffer, 2019; Guy and McCandless, 2012, 2020).

Equitable access to resources is a cornerstone of inclusive policy-making. This involves designing policies that ensure all individuals, regardless of socio-economic status, race, gender, or disability, have access to essential resources such as education, healthcare, and housing. For instance, educational policies might include scholarships and support programs for underrepresented students, ensuring they have the resources to succeed academically. Healthcare policies could focus on providing affordable and accessible health services to underserved communities, addressing disparities in health outcomes. Housing policies may include measures to ensure affordable housing options are available to low-income families and individuals.

Opportunities for economic participation are another critical aspect of inclusive policies. Employment policies should promote equal opportunities in the job market, including anti-discrimination measures, support for diverse hiring practices, and initiatives to improve workplace inclusion. Programs that offer vocational training and career development support can help individuals from marginalized groups gain the skills needed to compete in the job market. Policies encouraging entrepreneurship and providing access to funding for small businesses owned by underrepresented groups contribute to economic equity.

Inclusive policy-making also requires the involvement of diverse stakeholders in the decision-making process. Ensuring that policies reflect the needs and perspectives of all societal groups involves actively engaging with community members, advocacy organizations, and other relevant stakeholders. This participatory approach helps identify specific needs, address potential barriers, and create more effective and equitable solutions. Public consultations, community forums, and advisory committees can facilitate this engagement.

Addressing systemic barriers through inclusive policies is crucial for overcoming entrenched inequalities. This involves identifying and dismantling institutional biases and discriminatory practices that perpetuate inequality. For example, reforms in the criminal justice system to address racial profiling and disparities or changes in healthcare policies to ensure coverage for mental health services can help create more equitable systems. Additionally, promoting diversity and inclusion within policy-making bodies and institutions ensures that diverse perspectives are represented in the creation and implementation of policies.

Monitoring and evaluation are essential for assessing the effectiveness of inclusive policies. Regularly reviewing and analyzing policies' impacts helps ensure they achieve their intended goals and make a positive difference in addressing inequalities. Feedback mechanisms, such as surveys and impact assessments, can provide valuable insights into policies' functioning and where improvements may be needed (Shaeffer, 2019; Guy and McCandless, 2012, 2020). Inclusive policies are vital for ensuring equitable access to resources and opportunities for all societal groups. By focusing on equitable resource distribution, promoting economic participation, involving diverse stakeholders, addressing systemic barriers, and continuously monitoring and evaluating policies, societies can create a more just and inclusive environment where everyone has the opportunity to thrive.

4.2 Education and Empowerment

Enhance access to quality education and empower marginalized communities. Education and empowerment are critical to advancing social equity and fostering sustainable development. By enhancing access to quality education and empowering marginalized communities, societies can break the cycle of poverty, promote social mobility, and ensure that all individuals have the opportunity to contribute meaningfully to their communities and beyond Pranis, 2001; Burnett et al., 2018; Khoday and Perch, 2012).

Access to quality education is a fundamental right and a powerful tool for personal and societal advancement. Quality education gives individuals the skills and knowledge necessary to improve their circumstances and achieve their potential. However, marginalized communities often face barriers to accessing education, such as financial constraints, lack of infrastructure, and discrimination. To address these challenges, policies and programs should focus on reducing barriers to education. This includes providing financial support, such as scholarships and subsidies, to make education more affordable. Additionally, investing in educational infrastructure, such as building schools in underserved areas and providing necessary resources and materials, helps ensure that all students have access to a conducive learning environment.

Empowering marginalized communities involves creating opportunities and providing support that enables individuals and groups to take control of their lives and influence their surroundings. Empowerment can be achieved through various means, including education, skills training, and leadership development. Programs that offer vocational training, mentorship, and career development help individuals build the skills needed to enter the workforce and pursue economic opportunities. Furthermore, leadership and advocacy training can enable individuals from marginalized communities to actively participate in decision-making processes and advocate for their rights and interests.

Inclusive education ensures that all students receive a quality education regardless of their background or abilities. This involves implementing policies that support diverse learning needs, such as special education services for students with disabilities and culturally responsive teaching practices. By fostering an inclusive educational environment, schools can accommodate diverse learners and promote equity in educational outcomes.

Community involvement plays a significant role in both education and empowerment. Engaging community members in educational initiatives and decision-making processes helps ensure that programs are relevant and responsive to local needs. Community-based organizations can provide additional support, such as after-school programs and parental engagement activities, which enhance educational opportunities and outcomes.

Promoting gender equality in education and empowerment is also vital. Girls and women, particularly in marginalized communities, often face additional barriers to education and economic participation. Addressing these barriers through targeted initiatives, such as scholarships for girls, programs that challenge gender stereotypes, and support for women's entrepreneurship, helps create a more equitable environment where everyone has the chance to succeed.

Monitoring and evaluating educational and empowerment programs are essential for ensuring their effectiveness and making necessary adjustments. Regular assessments help identify gaps, measure progress, and ensure that interventions achieve their goals. Feedback from participants and stakeholders can provide valuable insights into how programs can be improved and tailored to better meet the needs of marginalized communities.

Enhancing access to quality education and empowering marginalized communities are key to fostering social equity and sustainable development. By addressing barriers to education, supporting skill-building and leadership, promoting inclusivity, and engaging communities, societies can create opportunities for all individuals to thrive and contribute to a more just and equitable world (Pranis, 2001; Burnett et al., 2018; Khoday and Perch, 2012).

4.3 Healthcare Access

Ensure universal access to healthcare services to improve overall well-being. Healthcare access is fundamental to improving overall well-being and ensuring everyone can lead healthy, productive lives. Ensuring universal access to healthcare involves removing barriers to care, addressing disparities, and creating a system that provides equitable and comprehensive health services to everyone, regardless of socio-economic status, geographic location, or personal circumstances (Peter, 2001; Hart, 1974; Johnson and Svara, 2011; Eden et al., 2024).

Universal access to healthcare means that every individual can receive essential health services without financial hardship. This involves a system where healthcare services are available to all, including preventive care, treatment for illnesses and injuries, and managing chronic conditions. To achieve this, policies must focus on expanding coverage and reducing individual costs (Peter, 2001; Hart, 1974; Johnson and Svara, 2011; Eden et al., 2024). This can be accomplished through public health insurance programs, subsidies for private insurance, or a universal healthcare system that funds care through general taxation. Ensuring that healthcare services are affordable and accessible helps prevent individuals from delaying or forgoing necessary care due to financial constraints.

Reducing disparities in healthcare access is critical for achieving equity. Marginalized and underserved populations often face significant barriers to accessing care, including geographic isolation, lack of transportation, and limited availability of services. Targeted initiatives such as mobile health clinics, telemedicine services, and community health programs can help bridge these gaps by bringing care to remote or underserved areas. Additionally, addressing social determinants of health—such as income, education, and housing—can improve overall access and reduce health disparities. For instance, policies that improve housing stability and food security can positively impact health outcomes and reduce the need for emergency medical services.

Quality of care is another essential component of healthcare access. Ensuring that healthcare services are available, high-quality, and patient-centered is crucial for improving health outcomes. This includes investing in healthcare infrastructure, such as modern facilities and equipment, and training healthcare professionals to provide effective and compassionate care. Implementing standards for care quality and patient safety helps ensure all individuals receive appropriate and effective treatments.

Preventive and primary care are vital for maintaining health and preventing the escalation of health issues. Universal access should include coverage for preventive services such as vaccinations, screenings, and health education. These services help detect and address health issues early, reducing the need for more complex and costly treatments later. Primary care is central in coordinating overall health and managing chronic conditions, promoting continuity of care and better health outcomes.

Health literacy and education are essential for empowering individuals to make informed decisions about their health. Providing resources and education on navigating the healthcare system, understanding health conditions, and following treatment plans enhances individuals' ability to manage their health effectively. Health literacy initiatives can help reduce misunderstandings and improve adherence to medical advice, contributing to better health outcomes.

Monitoring and evaluating healthcare access and services are essential for identifying gaps and areas for improvement. Regular assessments help measure the effectiveness of healthcare policies, track progress in reducing disparities, and ensure that resources are being used efficiently. Collecting healthcare utilization, outcomes, and patient satisfaction provides valuable insights for making evidence-based decisions and improving the healthcare system.

Ensuring universal access to healthcare services is crucial for enhancing overall well-being and achieving equitable health outcomes. By focusing on affordability, reducing disparities, maintaining quality, emphasizing preventive care, promoting health literacy, and continuously evaluating services, societies can create a healthcare system that meets the needs of all individuals and supports their right to health.

5

Technological Innovation

Technological innovation is crucial in advancing society by driving progress across various sectors, enhancing productivity, and improving quality of life. It encompasses the development and application of new technologies and methods that lead to significant improvements in processes, products, and services. Technological innovation is a key driver of economic growth and societal advancement, from breakthroughs in digital technology to biotechnology (Steil et al., 2002; Garud et al., 1997; Ochulor et al., 2024).

One of technological innovation's primary impacts is its ability to enhance efficiency and productivity. In industries such as manufacturing and agriculture, automation and robotics have revolutionized production processes, allowing for higher output and reduced labor costs. Artificial intelligence (AI) and machine learning have optimized operations by enabling data-driven decision-making and predictive analytics. These advancements not only increase productivity but also lead to the creation of new markets and job opportunities, driving economic growth and development.

Technological innovation also plays a vital role in addressing global challenges and improving quality of life. Innovations such as telemedicine, wearable health monitors, and advanced diagnostic tools have transformed patient care, making healthcare more accessible and personalized. Biotechnology advancements, including gene editing and personalized medicine, promise to treat and prevent diseases more effectively. Similarly, innovations in renewable energy technologies, such as solar panels and wind turbines, contribute to sustainable development by reducing reliance on fossil fuels and mitigating environmental impacts.

The role of technology in education has also expanded dramatically. E-learning platforms, digital resources, and interactive tools have made education more accessible and engaging. Technology enables remote learning and provides resources catering to diverse learning styles, reaching a broader audience and supporting lifelong learning.

However, technological innovation also presents challenges that must be addressed. Data privacy, cybersecurity, and the digital divide highlight the need for policies and practices that protect individuals and ensure equitable access to technology. As technology evolves, it is essential to address ethical considerations and ensure that advancements benefit society rather than exacerbate inequalities.

Technological innovation is a powerful force that drives progress and improves various aspects of life. By enhancing productivity, addressing global challenges, and transforming sectors such as healthcare and education, technology contributes to economic and social development. Balancing innovation with ethical considerations and ensuring equitable access is crucial for maximizing the benefits of technology and fostering a more inclusive and sustainable future (Steil et al., 2002; Garud et al., 1997; Ochulor et al., 2024).

5.1 Green Technology

Invest in and develop technologies that reduce environmental impact and promote sustainability. Green technology is developing and applying innovative technologies designed to reduce environmental impact and promote sustainability. This field encompasses a wide range of solutions to minimize the adverse effects of human activities on the environment while supporting long-term ecological health and resource efficiency. Investing in and advancing green technologies is essential for addressing pressing environmental challenges such as climate change, resource depletion, and pollution (Li et al., 2019; Mensah et al., 2019; Lee et al., 2014).

One key area of green technology is renewable energy. Technologies such as solar panels, wind turbines, and hydropower systems harness natural resources to generate energy with minimal environmental impact. Unlike fossil fuels, which release greenhouse gases and contribute to climate change, renewable energy sources produce little to no emissions during operation. Investing in these technologies helps reduce carbon footprints, supports energy security, and reduces dependence on finite resources.

Energy efficiency technologies are another important aspect of green technology. These innovations aim to reduce energy consumption while maintaining or improving performance. For example, advancements in energy-efficient appliances, lighting (such as LED technology), and building systems (including smart thermostats and high-performance insulation) contribute to significant reductions in energy use. By enhancing efficiency, these technologies help lower greenhouse gas emissions, reduce utility costs, and conserve resources.

Green technology also includes sustainable transportation solutions that aim to minimize environmental impact. For instance, electric vehicles (EVs) and hybrid vehicles produce fewer emissions than conventional gasoline-powered cars. Investments in EV infrastructure, such as charging stations, and developing alternative fuels, like hydrogen, further support the transition to cleaner transportation. Additionally, innovations in public transportation systems and non-motorized transport options, such as bicycles and pedestrian-friendly infrastructure, reduce urban air pollution and traffic congestion.

Waste management and recycling technologies are vital for reducing environmental impact. Technologies that enhance recycling processes, such as advanced sorting systems and chemical recycling methods, help reduce the amount of waste sent to landfills and promote the circular economy. Innovations in composting and waste-to-energy technologies also contribute to managing organic waste and recovering valuable resources.

Water conservation and treatment technologies are essential for addressing the growing challenges of water scarcity and pollution. Green technologies in this area include advanced water purification systems, efficient irrigation techniques, and water recycling and reuse methods. By improving water management and reducing waste, these technologies help ensure the availability of clean water and protect aquatic ecosystems.

Green technology investments also support research and development to drive further innovations and improve existing solutions. Public and private sector investments can accelerate the commercialization of green technologies and facilitate their adoption across various industries. Policymaking and incentives, such as subsidies for renewable energy projects and tax credits for energy-efficient upgrades, are crucial in promoting the widespread use of green technologies.

Green technology encompasses a range of innovations to reduce environmental impact and promote sustainability. By investing in renewable energy, enhancing energy efficiency, developing sustainable transportation, improving waste management, and advancing water conservation technologies, societies can address environmental challenges and support long-term ecological health. Supporting research, development, and adoption of these technologies is essential for creating a sustainable future and mitigating the effects of climate change and resource depletion (Li et al., 2019; Mensah et al., 2019; Lee et al., 2014).

5.2 Digital Transformation

Utilize digital tools and technologies to improve efficiency and reduce resource consumption. Digital transformation refers to integrating digital tools and technologies into various aspects of organizations and societal functions to enhance efficiency, improve processes, and reduce resource consumption. This transformative approach leverages technological advancements to drive innovation, streamline operations, and create more sustainable and resilient systems across industries (Davelaar and Nijkamp, 1990; Garud et al., 1997; Mehmood et al., 2024).

One significant impact of digital transformation is its ability to improve operational efficiency. Cloud computing, big data analytics, and the Internet of Things (IoT) enable organizations to optimize their operations by providing real-time insights and automating processes. For example, cloud-based platforms allow businesses to scale resources according to demand, reducing the need for physical infrastructure and associated energy consumption. IoT devices can monitor and control energy use in buildings and industrial processes, leading to more efficient resource management and lower operational costs.

Digital tools also facilitate better decision-making through advanced data analytics. Organizations can gain valuable insights into patterns, trends, and performance metrics by collecting and analyzing large volumes of data. This data-driven approach enables more informed decision-making, helping organizations identify opportunities for improvement, reduce waste, and implement more sustainable practices. For instance, predictive maintenance powered by data analytics can foresee equipment failures before they occur, reducing downtime and minimizing resource wastage.

Due to digital transformation, increased efficiency and resource savings are evident in various sectors. Digital technologies such as 3D printing and smart manufacturing systems enable more precise production, reducing material waste and energy consumption. In agriculture, precision farming techniques use data and digital tools to optimize planting, irrigation, and harvesting practices, leading to more efficient use of water and fertilizers. Similarly, digital platforms for remote work and virtual collaboration reduce the need for physical office space and travel, contributing to lower carbon footprints.

Digital transformation also promotes sustainability through improved supply chain management. Advanced digital tools allow for better tracking and optimizing supply chains, helping organizations reduce excess inventory, minimize transportation emissions, and streamline logistics. Blockchain technology, for example, can enhance transparency and traceability in supply chains, ensuring that products are sourced and manufactured sustainably.

Digital tools enhance consumer engagement and promote more sustainable behaviors. E-commerce platforms, for instance, can offer digital alternatives to physical products, reducing packaging waste and transportation emissions. Digital apps and platforms that support sustainable practices, such as energy management systems and recycling programs, empower individuals and organizations to make more environmentally conscious choices.

Digital transformation also presents challenges that need to be addressed. Data privacy, cybersecurity, and the digital divide must be managed to ensure that the benefits of digital tools are equitably distributed and that sensitive information is protected (Davelaar and Nijkamp, 1990; Garud et al., 1997; Mehmood et al., 2024). Additionally, the environmental impact of digital infrastructure, such as data centers and electronic waste, must be considered, and efforts should be made to minimize these impacts through energy-efficient designs and responsible recycling practices.

Digital transformation leverages digital tools and technologies to enhance efficiency and reduce resource consumption across various sectors. By integrating cloud computing, big data analytics, IoT, and other advanced technologies, organizations can streamline operations, improve decision-making, and promote sustainable practices. While challenges exist, the benefits of digital transformation in driving innovation and supporting environmental sustainability make it a crucial component of modern progress and efficiency.

5.3 Research and Development

Support research initiatives that address global challenges and drive innovation. Research and development (R&D) are pivotal in addressing global challenges and driving innovation across various fields. By investing in R&D, societies can develop new technologies, improve existing systems, and generate knowledge to solve pressing issues such as climate change, public health crises, and resource scarcity. Supporting R&D initiatives fosters economic growth, enhances the quality of life, and ensures sustainable development (Clow, 1998; Miremadi, 2019; Salikhova and Krekhivsky, 2018).

The primary role of R&D is to tackle global challenges through innovative solutions. For instance, in the field of climate change, research focuses on developing renewable energy technologies, improving energy efficiency, and understanding the impacts of environmental changes. Innovations such as advanced solar panels, wind turbines, and carbon capture technologies result from extensive R&D efforts to reduce greenhouse gas emissions and mitigate climate impacts. Similarly, research in agriculture leads to the development of sustainable farming practices and crops that are more resilient to changing climate conditions, helping to ensure food security.

Public health is another area where R&D plays a critical role. Research into new treatments, vaccines, and diagnostic tools helps combat diseases and improve health outcomes. The rapid development of COVID-19 vaccines is a prime example of how focused R&D can address urgent global health crises. Ongoing research into cancer, neurodegenerative diseases, and antimicrobial resistance is essential for advancing medical science and improving public health.

R&D also drives technological innovation, leading to advancements in various sectors. Research in information technology has resulted in breakthroughs such as artificial intelligence (AI), machine learning, and blockchain, which are transforming industries and creating new opportunities. In manufacturing, automation, and materials science innovations enhance production efficiency and enable the creation of new products. By supporting R&D in these areas, societies can foster technological progress that drives economic growth and enhances quality of life.

Sustainable development is another key focus of R&D. Research into sustainable practices and technologies aims to balance economic growth with environmental protection. Innovations in water management, waste reduction, and green technology are developed through R&D efforts to promote more sustainable resource use and reduce environmental impact. For example, advancements in water purification and recycling technologies help address water scarcity and ensure clean water access.

Collaboration and partnerships are essential for successful R&D. Public-private partnerships, academic collaborations, and international research networks enable sharing knowledge, resources, and expertise. Collaborative efforts help address complex challenges that require interdisciplinary approaches and global cooperation. For instance, addressing climate change or pandemics requires coordinated research efforts and the integration of diverse scientific disciplines.

Funding and support for R&D are critical for driving innovation and achieving meaningful outcomes. Governments, private sector organizations, and philanthropic institutions significantly provide financial resources and incentives for research initiatives. Grant programs, research institutions, and innovation hubs support scientists and researchers in pursuing their work and translating discoveries into practical applications.

Supporting research and development is crucial for addressing global challenges and driving innovation. By investing in R&D, societies can develop solutions to pressing issues such as climate change, public health, and resource management. Collaboration, funding, and a focus on sustainable development are essential to successful R&D efforts, leading to technological advancements, improved quality of life, and long-term societal benefits (Clow, 1998; Miremadi, 2019; Salikhova and Krekhivsky, 2018).

6

Circular Economy

The circular economy is an economic model that emphasizes the continual use of resources by promoting the reuse, repair, refurbishment, and recycling of products and materials. Unlike the traditional linear economy, which follows a 'take-make-dispose' pattern, the circular economy aims to create a closed-loop system where waste and resource input are minimized, and value is retained within the economy for as long as possible. This approach seeks to reduce environmental impact, conserve resources, and foster sustainability.

The concept of a circular economy revolves around several key principles. Design for longevity is one of these principles, where products are created with durability and repairability in mind. By focusing on designing products that last longer and are easier to maintain, the need for frequent replacements is reduced, thereby decreasing waste and resource consumption. Another principle is resource efficiency, which involves optimizing the use of materials and energy throughout the product lifecycle. This can be achieved through innovative manufacturing processes, such as using fewer materials or integrating renewable resources (Arruda et al., 2021; Kirchherr et al., 2017; Sumter et al., 2020).

A core aspect of the circular economy is product life extension. This involves extending the usable life of products through repair, refurbishment, and remanufacturing. Instead of discarding products at the end of their initial use, they are repaired or upgraded to maintain their functionality and value. This approach reduces waste and creates economic opportunities in the repair and refurbishment industries. For example, businesses specializing in refurbishing electronics or furniture help divert waste from landfills and provide consumers with more sustainable purchasing options.

Recycling and material recovery are also fundamental to the circular economy. By recycling materials from end-of-life products, valuable resources can be recovered and reintroduced into the production process. This reduces the demand for virgin materials and minimizes the environmental impact of resource extraction. Technologies and processes that improve recycling efficiency, such as advanced sorting systems and chemical recycling methods, are crucial for achieving effective material recovery.

Circular business models further exemplify this approach. These models include products-as-a-service, where businesses retain ownership of the product and provide it as a service to customers. For example, in a product-as-a-service model for electronics, a company might lease devices to customers, take responsibility for maintenance and upgrades, and reclaim the products at the end of their useful life for recycling or refurbishment. This shifts the focus from ownership to usage, aligning with circular economy principles and encouraging more sustainable consumption patterns.

Transitioning to a circular economy requires systemic change and collaboration across various sectors. Governments, businesses, and consumers all play a role in fostering this transition. Policies and regulations incentivizing circular practices, such as extended producer responsibility and recycling mandates, support shifting toward a more sustainable model. Businesses need to innovate and adopt circular strategies, while consumers can contribute by making informed choices and participating in recycling and repair programs.

The circular economy is a transformative approach that aims to create a sustainable and resource-efficient economic system. It seeks to minimize waste, conserve resources, and reduce environmental impact by focusing on product longevity, resource efficiency, life extension, and recycling (Arruda et al., 2021; Kirchherr et al., 2017; Sumter et al., 2020). Embracing circular principles requires a collaborative effort from all stakeholders and represents a significant shift toward more sustainable economic practices.

6.1 Resource Recycling

Promote the recycling and reuse of materials to minimize waste. Resource recycling is a critical strategy for minimizing waste and promoting sustainability by reprocessing used materials into new products. This process reduces the need for raw materials, lowers environmental impact, conserves energy, and reduces landfill use. By emphasizing the recycling and reuse of materials, societies can shift towards more sustainable consumption patterns and foster a circular economy (Gregson et al., 2015; Wainaina et al., 2020; Winterstetter et al., 2021).

The recycling process begins with the collection and sorting of recyclable materials. Effective recycling programs involve separating different materials, such as paper, glass, plastics, and metals, to ensure they are processed correctly. Advanced sorting technologies, such as automated systems and optical scanners, improve the efficiency and accuracy of this process, reducing contamination and increasing the quality of recycled materials.

Recycling offers numerous environmental benefits. It reduces the need for virgin raw materials, which decreases habitat destruction, mining, and resource extraction. For example, recycling aluminum saves up to 95% of the energy required to produce new aluminum from bauxite ore. Similarly, recycling paper reduces the need for deforestation and water use associated with new paper production. By diverting waste from landfills and incineration, recycling also helps lower greenhouse gas emissions and reduces the environmental impact of waste disposal.

Promoting the reuse of materials is closely related to recycling and involves finding new uses for products or materials instead of discarding them. Reuse can occur at various levels, from simple practices like repurposing containers to more complex ones like refurbishing electronics or furniture (Gregson et al., 2015; Wainaina et al., 2020; Winterstetter et al., 2021). This approach not only extends the lifecycle of products but also reduces the demand for new resources and minimizes waste. For example, repurposing glass jars for storage or donating used furniture to charities both contribute to waste reduction and support sustainability.

Public awareness and education are essential for effective recycling and reuse. Programs that inform individuals about what can be recycled, how to sort materials properly, and the benefits of recycling can significantly improve participation rates. Community initiatives, such as recycling drives and educational workshops, help raise awareness and encourage responsible behavior. Additionally, promoting the environmental and economic benefits of recycling reinforces its importance and motivates individuals to engage in sustainable practices.

Businesses and industries also play a crucial role in resource recycling. Companies can implement recycling programs within their operations, design products with recyclable materials, and invest in recycling technologies. By integrating recycling into their business models, companies reduce their environmental footprint and create value by recovering materials. For example, companies that use recycled plastics in their products contribute to reducing plastic waste and supporting circular supply chains.

Policy and regulation are vital for supporting resource recycling efforts. Governments can establish recycling mandates, provide incentives for recycling initiatives, and invest in recycling infrastructure. Policies encouraging extended producer responsibility (EPR) require manufacturers to take responsibility for their products' lifecycles, including end-of-life management. Such regulations drive industry-wide improvements in recycling practices and promote sustainable product design.

Resource recycling is a fundamental practice for minimizing waste and promoting sustainability. By focusing on efficient collection, sorting, and processing of recyclable materials and encouraging the reuse of products, societies can reduce environmental impact and conserve resources. Public awareness, business involvement, and supportive policies are essential for advancing recycling efforts and fostering a more sustainable future.

6.2 Product Lifecycle

Design products focusing on durability, repairability, and recyclability. Product lifecycle management involves designing and managing products, emphasizing their entire lifespan, from creation to disposal. By focusing on durability, repairability, and recyclability, manufacturers and designers can create functional, efficient, sustainable, and environmentally friendly products (Mestre and Cooper, 2017; Subramoniam et al., 2021; Diaz et al., 2021).

Durability is a crucial aspect of product lifecycle design. Products designed with durability are built to withstand wear and tear, extending their usable life and reducing the need for frequent replacements. This involves using high-quality materials, robust construction methods, and rigorous testing to ensure the product remains functional. Durable products contribute to sustainability by minimizing waste and resource consumption, as they do not need to be replaced as often. For example, well-designed appliances, furniture, and electronics that last longer help reduce the frequency of disposal and the demand for new products.

Repairability is another crucial element of product lifecycle management. Designing products that are easy to repair allows users to address issues and extend the product's life instead of discarding it when it malfunctions. This involves creating products with modular components that can be easily accessed and replaced, providing clear repair instructions, and offering readily available spare parts. Repairable products reduce waste and empower consumers to maintain and upgrade their items, promoting a culture of sustainability. For instance, many companies now design smartphones and laptops with removable batteries and modular parts, making it easier for users to repair or upgrade their devices.

Recyclability focuses on a product's end-of-life stage. Designing for recyclability involves using materials that can be efficiently recycled and ensuring that the product can be disassembled or processed without significant difficulty. This includes avoiding using mixed materials that are hard to separate or hazardous substances that complicate recycling. Products designed with recyclability in mind facilitate the recovery of valuable materials and reduce the environmental impact associated with waste. For example, many modern products are made from recyclable plastics or metals, and manufacturers often provide instructions on properly recycling their products.

Lifecycle thinking also considers the environmental impact throughout the entire product lifecycle, from raw material extraction and manufacturing to use and disposal. This approach helps identify opportunities for reducing environmental impact at every stage. For instance, using energy-efficient manufacturing processes and minimizing packaging can significantly reduce a product's carbon footprint. Additionally, designing products for easy disassembly allows for better separation and recycling of materials, further supporting sustainability goals.

Consumer education and involvement are essential for maximizing the benefits of durable, repairable, and recyclable products. Educating consumers about the advantages of these designs and how to maintain and dispose of products properly can enhance their engagement with sustainable practices. For example, providing guides on product care, repair, and recycling encourages consumers to make informed choices and extend the life of their products.

Designing products that focus on durability, repairability, and recyclability is essential for effective product lifecycle management and sustainability. Manufacturers and designers can significantly reduce environmental impact and resource consumption by creating products that last longer, are easier to repair, and can be efficiently recycled (Mestre and Cooper, 2017; Subramoniam et al., 2021; Diaz et al., 2021). This approach supports a shift towards more sustainable consumption patterns and contributes to a circular economy, where products and materials are continuously reused and regenerated.

6.3 Waste Management

Implement efficient waste management systems to reduce landfill use and pollution. Waste management is a crucial aspect of environmental stewardship. It aims to efficiently handle and process waste materials to minimize their impact on landfills and reduce pollution. Effective waste management systems are essential for protecting natural resources, reducing environmental hazards, and promoting sustainability. By implementing comprehensive waste management strategies, societies can manage waste more effectively and work towards a more sustainable future (Hidalgo et al., 2019; Tomić and Schneider, 2020; Malinauskaite et al., 2017).

Efficient waste management begins with waste reduction at the source. This involves designing processes, products, and systems that generate less waste and encourage sustainable consumption. Strategies for waste reduction include minimizing packaging, optimizing manufacturing processes, and promoting reusable materials. For example, businesses can reduce waste by adopting lean production techniques that eliminate inefficiencies and designing products with minimal or recyclable packaging.

Segregation and sorting are vital components of effective waste management. Waste can be processed more efficiently and effectively by separating waste into categories such as recyclables, organics, and hazardous materials. Implementing robust sorting systems at the source and waste processing facilities helps ensure materials are correctly classified and handled. Advanced sorting technologies, such as automated systems and conveyor belts equipped with sensors, enhance the accuracy and efficiency of waste segregation.

Recycling is a central element of waste management, focusing on processing used materials into new products to reduce the need for virgin resources. Efficient recycling systems involve collecting, sorting, and processing recyclable materials like paper, glass, plastics, and metals. By investing in recycling infrastructure and technologies, such as modern recycling facilities and advanced sorting equipment, communities can improve recycling rates and reduce the volume of waste sent to landfills.

Composting is another key strategy for managing organic waste. Composting involves decomposing organic materials, such as food scraps and yard waste, into nutrient-rich compost that can be used as a natural fertilizer. Implementing community composting programs and promoting backyard composting can help divert organic waste from landfills and support soil health and sustainability. Effective composting systems require proper management of composting conditions, such as moisture levels and aeration, to ensure efficient decomposition and high-quality compost.

Waste-to-energy technologies convert waste into energy, providing an alternative approach to waste management. Technologies like incineration and anaerobic digestion can generate electricity or heat from waste materials. While these technologies can reduce the volume of waste and produce valuable energy, managing emissions and residues responsibly is crucial. Modern waste-to-energy facilities are designed to minimize environmental impacts through advanced emission controls and efficient energy recovery systems.

Education and public engagement are essential for successful waste management. Educating residents and businesses about waste reduction, recycling, and composting helps foster responsible waste management practices and increases participation in waste diversion programs. Public awareness campaigns, community workshops, and clear guidelines for waste separation can enhance understanding and compliance with waste management practices.

Policy and regulation play a significant role in shaping waste management systems. Governments can implement policies and regulations that promote waste reduction, recycling, and proper disposal. Measures such as extended producer responsibility (EPR) require manufacturers to take responsibility for the lifecycle of their products, including waste management (Hidalgo et al., 2019; Tomić and Schneider, 2020; Malinauskaite et al., 2017). Incentives, penalties, and regulations can drive improvements in waste management practices and support the development of sustainable waste management infrastructure.

Implementing efficient waste management systems is crucial for reducing landfill use and pollution. By focusing on waste reduction, segregation, recycling, composting, and waste-to-energy technologies, societies can manage waste more effectively and minimize environmental impact. Education, public engagement, and supportive policies are essential for enhancing waste management practices and promoting sustainability. Through these efforts, communities can work towards a more sustainable and resource-efficient future.

7

Global Cooperation

Global cooperation refers to the collaborative efforts of nations, organizations, and individuals to address issues that transcend national borders and require collective action. It involves working together to tackle global challenges, share resources and knowledge, and promote common goals for the benefit of the international community. This collaborative approach is essential for addressing complex, interconnected problems that no single entity can solve, such as climate change, pandemics, and economic inequality (Woodward, 2007; Nolan, 2010; Wagner et al., 2015).

The meaning of global cooperation lies in its capacity to harness the collective strengths and resources of diverse actors to achieve shared objectives. It encompasses various forms of collaboration, including bilateral agreements between countries, multilateral agreements involving multiple nations, and partnerships between governments, international organizations, non-governmental organizations (NGOs), and the private sector. Through these collaborative efforts, stakeholders can pool expertise, exchange best practices, and coordinate actions to address global issues more effectively (Bhandari and Adhikari, 2024).

The importance of global cooperation is evident in several critical areas. First and foremost, global challenges such as climate change, environmental degradation, and public health crises require a unified response. For instance, climate change affects ecosystems, weather patterns, and economies worldwide, necessitating coordinated efforts to reduce greenhouse gas emissions, transition to renewable energy, and adapt to changing conditions. Without global cooperation, individual nations' efforts may be insufficient or counterproductive, undermining progress and exacerbating the problem.

Public health is another area where global cooperation is crucial. The COVID-19 pandemic demonstrated the need for international collaboration in managing health crises. Coordinated efforts in vaccine development, distribution, and sharing of medical resources were essential for controlling the spread of the virus and mitigating its impact. Global cooperation also plays a vital role in addressing other health issues, such as combating diseases like malaria and HIV/AIDS, ensuring access to healthcare, and improving global health outcomes.

Economic stability is closely linked to global cooperation. Economic interdependence means financial instability or economic downturns in one region can have ripple effects worldwide. Collaborative efforts to address economic challenges, such as trade imbalances, financial crises, and poverty, help promote global economic stability and development. International organizations like the International Monetary Fund (IMF) and the World Bank facilitate cooperation and support economic policies that benefit the global economy.

Global cooperation fosters peace and security by promoting dialogue and collaboration among nations. International treaties, peacekeeping missions, and conflict resolution efforts exemplify how cooperative approaches can address geopolitical tensions and prevent conflicts. By working together, countries can address the root causes of conflicts, promote mutual understanding, and build frameworks for peaceful coexistence.

Global cooperation supports sustainable development by advancing shared goals related to poverty reduction, education, and environmental sustainability. Initiatives such as the United Nations Sustainable Development Goals (SDGs) highlight the importance of collective action in achieving these objectives. Countries and organizations can progress toward a more equitable and sustainable future by collaborating on projects and sharing resources (Woodward, 2007; Nolan, 2010; Wagner et al., 2015).

Global cooperation involves collective action by nations, organizations, and individuals to address transnational issues and achieve shared goals. Its importance is evident in tackling global challenges such as climate change, public health crises, economic instability, and peace and security. By working together, stakeholders can leverage their collective strengths, pool resources, and coordinate efforts to create a more sustainable, equitable, and stable world.

7.1 International Agreements

Strengthen global treaties and agreements on climate change, biodiversity, and sustainable development. International agreements are vital tools for addressing global challenges that no nation can resolve alone. By strengthening global treaties and agreements on climate change, biodiversity, and sustainable development, the international community can create a framework for collective action, ensure compliance with shared goals, and foster cooperation among countries. These agreements help set common standards, provide mechanisms for accountability, and mobilize resources to tackle pressing global issues (Barrett, 1999; Paulo, 2014; Georghiou, 1998).

Strengthening global treaties on climate change is essential for mitigating the impacts of global warming and promoting sustainable development. The Paris Agreement, adopted in 2015, is a landmark international treaty that aims to limit global temperature rise to well below 2 degrees Celsius above pre-industrial levels, with efforts to limit the increase to 1.5 degrees Celsius. Strengthening this agreement involves enhancing national commitments, known as Nationally Determined Contributions (NDCs), to reduce greenhouse gas emissions. It also requires improving transparency and accountability mechanisms to ensure that countries meet their targets. Additionally, providing financial and technical support to developing countries helps them transition to low-carbon economies and adapt to climate impacts, fostering global solidarity and equitable progress.

Biodiversity conservation is another critical area where international agreements play a pivotal role. The Convention on Biological Diversity (CBD), established in 1992, aims to conserve biological diversity, promote sustainable use of its components, and ensure fair and equitable sharing of benefits arising from genetic resources. Strengthening the CBD involves setting more ambitious targets for biodiversity protection, enhancing the implementation of conservation measures, and increasing funding for biodiversity initiatives. Collaborative efforts, such as transboundary conservation projects and global wildlife protection programs, are essential for preserving multiple countries' ecosystems and species. By reinforcing these agreements, the international community can address threats to biodiversity, such as habitat loss, pollution, and climate change (Bhandari, 2022).

Sustainable development requires a holistic approach integrating economic, social, and environmental dimensions. The United Nations 2030 Agenda for Sustainable Development, adopted in 2015, provides a comprehensive framework with 17 Sustainable Development Goals (SDGs) that address many global challenges, including poverty, hunger, health, education, gender equality, clean water, and climate action. Strengthening international agreements on sustainable development involves aligning national policies and strategies with the SDGs, fostering partnerships between governments, businesses, and civil society, and enhancing monitoring and reporting mechanisms to track progress. International cooperation and resource mobilization are crucial for achieving the SDGs, especially in developing countries that face significant challenges in meeting these goals (Bhandari, 2024).

International agreements also promote cooperation and capacity-building among nations. Countries can share knowledge, expertise, and best practices by participating in global treaties. For example, international forums and conferences provide platforms for exchanging experiences and innovations in renewable energy, sustainable agriculture, and environmental conservation. Funded by international organizations and donor countries, capacity-building initiatives help developing nations build the institutional and technical capacity to implement sustainable development measures effectively.

Enforcement and compliance mechanisms are critical for the success of international agreements. Strengthening these mechanisms involves establishing clear guidelines, monitoring progress, and holding countries accountable for their commitments. International bodies, such as the United Nations Framework Convention on Climate Change (UNFCCC) and the CBD, play a key role in overseeing the implementation of agreements and facilitating negotiations among parties. Ensuring that countries adhere to their obligations helps maintain the credibility and effectiveness of international treaties.

Strengthening global treaties and agreements on climate change, biodiversity, and sustainable development is essential for addressing global challenges and promoting a sustainable future. These agreements provide a framework for collective action and ensure progress towards shared global goals by enhancing commitments, improving accountability, and fostering international cooperation (Barrett, 1999; Paulo, 2014; Georghiou, 1998). The international community can create a more sustainable, equitable, and resilient world through collaborative efforts and resource mobilization.

7.2 Shared Knowledge

Facilitate the exchange of knowledge and best practices across countries and organizations. Shared knowledge systematically exchanges information, expertise, and best practices across countries and organizations to foster learning, innovation, and progress. Facilitating such exchanges is crucial for addressing global challenges, improving efficiencies, and promoting sustainable development. When nations and organizations share their experiences and insights, they can collectively benefit from each other's successes and avoid repeating past mistakes (Derichs, 2017; Kuramoto and Sagasti, 2002; Petrillo and Bellaviti, 2018).

Facilitating the exchange of knowledge starts with creating platforms for dialogue and collaboration. International conferences, workshops, and forums allow stakeholders from various sectors to come together, discuss common issues, and share solutions. These events encourage networking, build partnerships, and foster an environment of mutual learning. For example, the annual United Nations Climate Change Conference (COP) brings together policymakers, scientists, and activists to share progress on climate action and negotiate new commitments.

Technology and digital platforms play a significant role in knowledge sharing. Online repositories, databases, and collaborative platforms enable the dissemination of research findings, policy briefs, and case studies across borders. Websites like the World Bank's Open Knowledge Repository and the United Nations' Sustainable Development Knowledge Platform provide free access to information on best practices, development projects, and policy recommendations. These resources help practitioners and policymakers stay informed about the latest developments and apply proven strategies in their contexts.

Educational exchanges and training programs are effective ways to transfer knowledge and skills. Scholarships, fellowships, and exchange programs allow students, researchers, and professionals to study and work in different countries, gaining exposure to diverse perspectives and practices. For example, the Fulbright Program facilitates academic exchanges that promote mutual understanding and collaboration between the United States and other countries. Similarly, the Erasmus+ program in the European Union supports education, training, youth, and sports exchanges, enhancing the skills and employability of participants.

Collaborative research and development (R&D) projects are another critical knowledge-sharing avenue. Joint research initiatives between institutions, countries, and organizations can address global challenges by pooling resources, expertise, and funding. International consortia, such as the Human Genome Project and the International Thermonuclear Experimental Reactor (ITER) project, demonstrate how collaborative R&D can lead to significant scientific breakthroughs and technological advancements. These projects enable the sharing of knowledge and the development of innovative solutions that no single entity could achieve alone.

Best practice dissemination involves documenting and sharing successful approaches to common challenges. Case studies, guidelines, and toolkits that outline effective strategies and methodologies can be distributed to a wide audience, helping others replicate and adapt these practices. For instance, the United Nations Food and Agriculture Organization (FAO) publishes best practice documents on sustainable agriculture, fisheries, and forestry, providing valuable insights for practitioners worldwide.

Capacity-building initiatives are essential for enabling knowledge transfer. Training workshops, seminars, and technical assistance programs help build the skills and capabilities of individuals and institutions, allowing them to apply new knowledge effectively. For example, the United Nations

Development Program (UNDP) supports capacity-building projects that enhance governance, resilience, and sustainable development in developing countries. These initiatives strengthen local capacities and ensure knowledge sharing leads to tangible and lasting improvements.

Policy dialogues and peer learning networks facilitate knowledge exchange among governments and policymakers. Forums like the Organization for Economic Co-operation and Development (OECD) provide platforms for countries to share policy experiences, conduct peer reviews, and develop best practice recommendations. These dialogues help harmonize policies, promote effective governance, and foster international cooperation on common issues.

Facilitating the exchange of knowledge and best practices across countries and organizations is crucial for addressing global challenges and promoting sustainable development. The international community can share insights, innovate, and collectively progress toward a more sustainable and equitable world through platforms for dialogue, digital resources, educational exchanges, collaborative R&D, best practice dissemination, capacity-building initiatives, and policy dialogues (Derichs, 2017; Kuramoto and Sagasti, 2002; Petrillo and Bellaviti, 2018).

7.3 Collaborative Efforts

To address global challenges, foster partnerships between governments, businesses, and civil society. Collaborative efforts are essential for addressing global challenges that are too complex for any single entity to tackle alone. The international community can pool resources, expertise, and influence to develop comprehensive and effective solutions by fostering partnerships between governments, businesses, and civil society. These collaborative efforts leverage the strengths of each sector to address issues such as climate change, poverty, and health crises, ultimately promoting sustainable development and global well-being (Rosenberg, 2010; Waham et al., 2023; Enarson, 2000).

Governments play a crucial role in collaborative efforts by providing policy frameworks, regulatory support, and funding. They can facilitate partnerships through legislation encouraging collaboration and creating public-private partnership (PPP) frameworks. For example, governments can incentivize renewable energy projects by offering tax breaks or subsidies to businesses that invest in green technologies (Bhandari, 2022). Moreover, governments can coordinate large-scale initiatives, such as national health campaigns or climate action plans, bringing together various stakeholders to work towards common goals.

Businesses bring innovation, investment, and operational expertise to collaborative efforts. Companies have the resources and capabilities to develop and implement new technologies, products, and services that address global challenges. By partnering with governments and civil society, businesses can align their operations with broader societal goals and contribute to sustainable development. For instance, companies in the tech industry can collaborate with educational institutions and NGOs to improve digital literacy and access to technology in underserved communities. Such partnerships not only benefit society but also create new markets and opportunities for businesses.

Civil society organizations (CSOs), including NGOs, advocacy groups, and community organizations, play a critical role in representing the interests of diverse populations and driving grassroots initiatives. These organizations often have deep connections within communities and can mobilize public support for collaborative efforts. CSOs can also provide valuable insights into the needs and preferences of local populations, ensuring that solutions are culturally appropriate and widely accepted. For example, environmental NGOs can work with local communities, governments, and businesses to promote conservation efforts and sustainable land use practices (Bhandari, 2020).

Effective collaboration requires clear communication, shared goals, and mutual trust. Establishing transparent decision-making, resource allocation, and accountability mechanisms helps ensure that all parties are committed to the partnership and its objectives. Regular communication and joint planning sessions allow partners to align their efforts, address challenges, and adapt strategies as needed. Collaborative efforts can achieve greater impact and sustainability by fostering a culture of cooperation and mutual respect.

Successful examples of collaborative efforts include multi-stakeholder initiatives such as the Global Alliance for Vaccines and Immunization (Gavi). Gavi brings together governments, international organizations, the private sector, and civil society to improve vaccine access in developing countries (Gandhi, 2015; Pagliusi et al., 2019). This partnership has successfully immunized millions of children, preventing numerous diseases and saving countless lives. Similarly, the United Nations Global Compact encourages businesses worldwide to adopt sustainable and socially responsible policies, fostering collaboration between companies, governments, and civil society to achieve the UN Sustainable Development Goals (SDGs).

Initiatives like the Carbon Disclosure Project (CDP) demonstrate the power of collaborative efforts in climate action (Matisoff et al., 2013). The CDP provides a platform for companies, cities, and regions to disclose their environmental impact and share strategies for reducing carbon emissions. By facilitating transparency and knowledge sharing, the CDP enables stakeholders to learn from each other and collectively advance climate action.

Education and awareness campaigns are another area where collaborative efforts can significantly impact. Partnerships between governments, businesses, and civil society can promote educational initiatives that raise awareness about critical issues such as climate change, health, and social justice (Gandhi, 2015; Pagliusi et al., 2019; Matisoff et al., 2013; Rosenberg, 2010; Waham et al., 2023; Enarson, 2000). For example, collaborations between schools, tech companies, and NGOs can enhance digital education, equipping students with the skills needed for the future job market while addressing digital divides.

Fostering partnerships between governments, businesses, and civil society is essential for effectively addressing global challenges. These collaborative efforts leverage the unique strengths of each sector, from policy support and innovation to community engagement and grassroots mobilization. These diverse stakeholders can develop comprehensive solutions, drive sustainable development, and create a more equitable and resilient world (Rosenberg, 2010; Waham et al., 2023; Enarson, 2000).

8

Community Engagement

A community is a group of individuals who share common interests, values, or geographical locations and often engage in regular interactions and activities. Communities can be based on various factors such as neighborhoods, cultural or religious affiliations, professional or academic interests, and social or recreational activities. The sense of belonging and mutual support among members is a fundamental characteristic of a community, fostering social cohesion and collective well-being (O'Mara et al., 2013, 2015; Shalowitz et al., 2009).

Community engagement involves members in decision-making, planning, and action on issues affecting their lives. It emphasizes the importance of active participation, collaboration, and empowerment, enabling individuals to contribute their knowledge, skills, and perspectives to address shared challenges and opportunities. Community engagement is crucial for building strong, resilient, and sustainable communities.

Community engagement is important because it fosters a sense of ownership and responsibility among community members. When people are actively involved in the processes that shape their environment and well-being, they are more likely to take pride in their community and work towards its improvement. This sense of ownership can lead to greater civic participation, such as volunteering, voting, and involvement in local governance, which strengthens democratic processes and promotes social cohesion.

Community engagement also enhances the effectiveness of policies and programs. By involving community members in the planning and implementation stages, policymakers and organizations can ensure that initiatives are tailored to the community's specific needs and preferences. This participatory approach increases the likelihood of successful outcomes and sustainability by leveraging local knowledge and fostering community buy-in. For example, community-led health initiatives often achieve better results because they are designed with an understanding of local contexts and cultural sensitivities.

Empowerment is another key benefit of community engagement. Community engagement fosters a sense of agency and empowerment by providing opportunities for individuals to contribute to decision-making processes. This can be particularly important for marginalized or disadvantaged groups, who may otherwise have limited opportunities to influence decisions that affect their lives. Empowered community members are more likely to advocate for their rights, seek out resources, and collaborate to address common challenges, leading to improved social and economic outcomes.

Social capital, the networks of relationships and trust within a community, is strengthened through engagement. When community members work together towards common goals, they build stronger connections and trust, which can enhance social support, reduce social isolation, and increase overall community resilience. These networks can be vital during times of crisis, such as natural disasters or economic downturns, as they enable communities to mobilize resources and support more effectively.

In addition, community engagement promotes innovation and creativity. Diverse perspectives and ideas emerge when people from different backgrounds and experiences come together to address common issues. This collaborative environment can lead to innovative solutions that might not have been possible through top-down approaches. For instance, community-driven urban planning initiatives often result in more inclusive and sustainable designs that reflect the unique needs and aspirations of the residents.

Economic benefits also arise from community engagement. Engaged communities can attract investment and resources by demonstrating a commitment to collective well-being and sustainability. As a result of increased community support and participation in economic activities, local businesses may thrive. Furthermore, community-led initiatives can create job opportunities and stimulate local economies by leveraging local skills and resources.

Community engagement is vital for building strong, resilient, and sustainable communities. It fosters a sense of ownership and responsibility, enhances the effectiveness of policies and programs, empowers individuals, strengthens social capital, promotes innovation, and delivers economic benefits (O'Mara et al., 2013, 2015; Shalowitz et al., 2009). By actively involving community members in decision-making and action, societies can create more inclusive, responsive, and vibrant communities better equipped to address challenges and seize opportunities.

8.1 Local Initiatives

Encourage community-led sustainability projects and local decision-making. Local initiatives are grassroots efforts driven by community members to address specific issues within their locality, often focusing on sustainability and community well-being. Encouraging community-led sustainability projects and local decision-making is crucial for fostering resilient, adaptive, and self-sufficient communities. These initiatives leverage local knowledge, resources, and leadership to create tailored solutions that meet the unique needs and aspirations of the community (Fischer et al., 2017).

Encouraging community-led sustainability projects involves empowering residents to take charge of their environmental and social goals. These projects can include community gardens, renewable energy cooperatives, local recycling programs, and sustainable transportation initiatives. By engaging community members in planning and implementing these projects, they become active participants in creating a more sustainable future. For example, community gardens provide fresh produce, foster community, improve local biodiversity, and offer educational opportunities about sustainable agriculture.

Local decision-making is essential for ensuring that the voices and needs of community members are heard and addressed. When decisions about local development, resource management, and public services are made at the community level, they are more likely to reflect the priorities and values of the residents. This localized approach can lead to more effective and equitable outcomes as it considers the community's specific context and circumstances. For instance, local decision-making in urban planning can ensure that development projects enhance public spaces, preserve cultural heritage, and promote social inclusivity.

Community-led initiatives often result in higher levels of engagement and commitment. When people feel a sense of ownership and responsibility for local projects, they are more likely to contribute their time, skills, and resources. This collective effort not only strengthens social bonds but also ensures the sustainability of the projects. Successful examples include neighborhood clean-up campaigns, local food cooperatives, and energy-saving programs, where community involvement has led to sustained positive impacts.

Empowerment through local initiatives fosters a sense of agency and confidence among community members. When individuals are involved in decision-making processes and see the tangible results of their efforts, they are more likely to believe in their ability to effect change. This empowerment can extend beyond the immediate project, inspiring community members to engage in other civic activities and advocate for broader social and environmental issues. For instance, a community that successfully manages a local water conservation project might be motivated to tackle other challenges, such as renewable energy adoption or waste reduction.

Local initiatives also encourage innovation and adaptability. Community members often possess deep knowledge of their local environment and can develop creative solutions uniquely suited to their context. This localized innovation can lead to more effective and sustainable practices that might not emerge from top-down approaches. For example, a coastal community might develop unique methods for protecting against sea level rise based on its specific geography and local expertise.

Community-led projects can yield economic benefits. By utilizing local resources and skills, these initiatives can create job opportunities, stimulate local economies, and reduce reliance on external resources. For instance, a local renewable energy project can generate employment for technicians, installers, and maintenance workers while reducing the community's energy costs. Additionally, local food production through community gardens or urban farms can enhance food security and support local markets.

Policy support and resources from governments and organizations are crucial for the success of local initiatives. Providing grants, technical assistance, and training can help communities build the capacity to plan and implement their projects effectively. Policies that decentralize decision-making and encourage community participation can create an enabling environment for local initiatives to thrive. For example, municipal governments can establish participatory budgeting processes, allowing residents to allocate funds to the most important projects.

Encouraging community-led sustainability projects and local decision-making is vital for fostering resilient, empowered, and adaptive communities. These initiatives leverage local knowledge and resources, promote engagement and commitment, empower residents, drive innovation, and generate economic benefits. By supporting and facilitating local initiatives, societies can create more sustainable and equitable communities better equipped to address their unique challenges and seize opportunities.

8.2 Public Awareness

Increase public awareness and education on environmental issues and sustainable practices. Public awareness refers to the knowledge and understanding that the general population has about various issues, in this case, environmental concerns and sustainable practices. Increasing public awareness and education on these topics is crucial for fostering a society that values and actively participates in environmental stewardship and sustainability. This approach ensures that individuals are informed and equipped to make responsible choices contributing to the planet's well-being (Khatibi et al., 2021; Aitken et al., 2016; Cattapan et al., 2020).

Public awareness and education on environmental issues cannot be overstated. When people are aware of the environmental challenges we face, such as climate change, deforestation, pollution, and biodiversity loss, they are more likely to support and engage in actions that mitigate these problems. Education helps bridge the gap between awareness and action by providing individuals with the knowledge and tools they need to adopt sustainable practices in their daily lives. For example, understanding the impact of single-use plastics can lead to increased use of reusable bags, bottles, and containers.

Educational campaigns and programs are effective tools for raising public awareness. These initiatives can take many forms, including school curricula, public service announcements, workshops, seminars, and community events. Incorporating environmental education into school programs from an early age helps cultivate a generation of environmentally conscious individuals. Schools can implement hands-on learning experiences, such as school gardens and recycling programs, to teach students about sustainability practically and engagingly. Public service announcements on television, radio, and social media can also reach a wide audience, delivering important messages about environmental conservation and sustainable living.

Community-based education initiatives play a significant role in increasing public awareness. Local organizations, NGOs, and community groups can organize clean-up drives, tree-planting activities, and sustainability fairs to engage residents and foster a sense of community responsibility. These activities raise awareness and provide tangible opportunities for individuals to contribute to environmental conservation efforts. For instance, community-led recycling programs can educate residents on proper waste segregation and recycling techniques, leading to reduced landfill waste and increased recycling rates.

Partnerships between governments, businesses, and civil society are essential for amplifying public awareness efforts. Governments can implement policies and regulations that promote environmental education and awareness, such as mandating eco-labeling on products, supporting green certifications, and funding environmental education programs. Businesses can play their part by adopting and promoting sustainable practices, offering eco-friendly products, and running awareness campaigns that educate consumers about the environmental benefits of their choices. Civil society organizations can advocate for environmental causes, mobilize volunteers, and create platforms for information sharing and community engagement.

Using digital platforms and social media is increasingly important for raising public awareness. The Internet and social media provide powerful tools for reaching diverse audiences and disseminating information quickly and widely. Environmental organizations and influencers can use these platforms to share educational content, promote sustainable practices, and inspire action. Campaigns such as Earth Hour, where people are encouraged to turn off their lights for one hour to raise awareness about climate change, leverage social media to engage millions of participants globally.

Creating compelling narratives and visual content can significantly enhance public engagement with environmental issues. Documentaries, films, and visual campaigns that highlight nature's beauty, environmental degradation's impacts, and the benefits of sustainable living can evoke emotional responses and motivate individuals to take action. For example, documentaries like "An Inconvenient Truth" and "Our Planet" have played pivotal roles in raising global awareness about climate change and the importance of conservation.

Encouraging personal responsibility and behavior change is key to public awareness campaigns. Providing practical tips and guidelines on how individuals can reduce their environmental footprint, such as conserving water, reducing energy consumption, and adopting a plant-based diet, can empower people to make sustainable choices. Additionally, showcasing success stories and positive examples of communities and individuals making a difference can inspire others to follow suit.

Increasing public awareness and education on environmental issues and sustainable practices is vital for fostering a society committed to environmental stewardship. Educational campaigns, community-based initiatives, partnerships, digital platforms, compelling narratives, and practical guidelines contribute to a more informed and engaged public (Khatibi et al., 2021; Aitken et al., 2016; Cattapan et al., 2020). Society can drive collective action towards a more sustainable and resilient future by empowering individuals with knowledge and tools.

8.3 Participatory Governance

Involve communities in planning and implementing sustainability policies. Participatory governance, a democratic approach, involves communities in the decision-making processes for planning and implementing sustainability policies. This governance model places a significant emphasis on citizens' active engagement, ensuring their voices are not only heard but also considered in shaping policies that directly affect their lives. The crucial role of communities in participatory governance is evident in the creation of compelling and inclusive sustainability policies that reflect the needs and aspirations of the community (Osborne et al., 2021; Suarez, 2024; Bhandari, 2024).

The importance of participatory governance lies in its ability to enhance transparency, accountability, and legitimacy in the policy-making process. When involved in planning and implementation, communities can contribute their local knowledge and expertise, leading to more informed and context-specific policies. This involvement fosters a sense of ownership and responsibility among community members, increasing their commitment to the success and sustainability of the initiatives. For example, in urban planning, involving residents in the design and development of public spaces can ensure that these areas meet the community's needs and promote social cohesion.

Community engagement in policy planning can take various forms, including public consultations, participatory budgeting, and stakeholder workshops. Public consultations allow citizens to express their views and concerns about proposed policies or projects. These consultations can be conducted through town hall meetings, focus groups, or online forums, allowing for diverse input from different community segments. For instance, in developing a local climate action plan, public consultations can gather insights into the community's priorities, such as energy efficiency, renewable energy adoption, or green infrastructure.

Participatory budgeting is a powerful tool for involving communities in decision-making about the allocation of public funds. This process allows citizens to propose and vote on projects they believe will benefit their community, ensuring that public resources are used effectively and equitably. Participatory budgeting has been successfully implemented in cities worldwide, leading to funding community-driven projects such as parks, playgrounds, and local business initiatives. Participatory budgeting fosters greater trust in local government and encourages active civic engagement by giving residents a direct say in spending public money.

Stakeholder workshops and collaborative planning sessions unite community members, policymakers, experts, and other stakeholders to co-create solutions to sustainability challenges. These workshops provide a space for dialogue, knowledge sharing, and joint problem-solving, enabling the development of innovative and practical policies. For example, in addressing water scarcity, a stakeholder workshop might involve local farmers, environmentalists, engineers, and government officials working together to design efficient water management strategies that balance agricultural needs with environmental conservation.

The benefits of participatory governance extend beyond policy development to the implementation phase. When communities are involved in implementing sustainability policies, they can help monitor progress, identify challenges, and suggest adjustments based on their lived experiences. This continuous feedback loop ensures that policies remain relevant and effective over time. For instance, community involvement in implementing waste management programs can lead to higher participation rates in recycling and composting initiatives, as residents feel a sense of ownership and responsibility for the program's success.

Building capacity for effective participation is essential for the success of participatory governance. This involves providing education and training to community members on the policy issues and the available participatory processes. Empowering citizens with the knowledge and skills needed to engage meaningfully in governance can enhance the quality of their contributions and ensure that diverse voices are heard. Governments and organizations can offer workshops, informational materials, and technical assistance to support community participation.

Ensuring inclusivity in participatory governance is crucial for achieving equitable outcomes. Special efforts must be made to engage marginalized and underrepresented groups, such as low-income residents, indigenous communities, and people with disabilities. Inclusive, participatory processes can be designed by providing accessible meeting venues, offering translation services, and using diverse communication channels to reach different community segments. Participatory governance can help address social inequalities and promote social justice by prioritizing inclusivity.

Participatory governance is essential for the effective planning and implementation of sustainability policies. This approach enhances transparency, accountability, and legitimacy by involving communities in decision-making, leading to more informed and context-specific policies. Tools such as public consultations, participatory budgeting, and stakeholder workshops facilitate community engagement while building capacity and ensuring inclusivity, which is key to the success of participatory governance (Osborne et al., 2021; Suarez, 2024; Bhandari, 2024). Through active community involvement, societies can develop and implement sustainability policies that are more responsive, equitable, and resilient.

9

Economic Transformation

Economic transformation refers to a comprehensive and systemic change in an economy's structure and functioning to achieve sustainable development, inclusivity, and long-term resilience. This process involves shifting from traditional economic activities, often reliant on finite resources and outdated practices, to more innovative, sustainable, and diversified economic models. Economic transformation aims to improve productivity, enhance economic diversification, and ensure equitable growth that benefits all segments of society (De Vries et al., 2021; Ahrens, 2006; Dolan, 1993).

9.1 The Key Components of Economic Transformation

Economic transformation for sustainable development involves several key components. Diversification is essential, expanding beyond traditional sectors into technology, green industries, and services to reduce dependency and increase resilience. Innovation is crucial in driving advancements in renewable energy and eco-friendly technologies. Inclusive growth ensures equitable opportunities for all, addressing inequality and promoting gender equality. Environmental sustainability is prioritized by adopting low-carbon, resource-efficient practices. Lastly, strong governance is vital, with transparent institutions and policies that support long-term, sustainable growth aligned with global sustainability goals, fostering a more resilient and equitable economy.

9.1.1 Industrial diversification: A path to economic resilience

Industrial diversification refers to the strategic broadening of a nation's or region's economic activities, moving away from reliance on a limited number of sectors, such as agriculture, mining, or other extractive industries. The goal is to develop a more varied economy that includes sectors such as manufacturing, services, technology, and creative industries. This shift enhances economic resilience by reducing vulnerability to sector-specific shocks, such as fluctuations in commodity prices or natural disasters, which can devastate economies heavily reliant on a single industry.

One of the key advantages of industrial diversification is the ability to stabilize economic growth. Economies centered on agriculture or extractive industries often face volatility due to external factors like market demand, climate change, or resource depletion. Expanding into other sectors, such as technology, services, and manufacturing, a country or region can mitigate the risks associated with over-reliance on these unpredictable industries. For example, suppose an agricultural sector faces a downturn due to adverse weather conditions. In that case, a well-diversified economy can rely on its growing technology or manufacturing sectors to maintain economic momentum and safeguard livelihoods.

Diversification also opens up opportunities for innovation and value creation. Economies moving from traditional industries to more advanced sectors can foster research and development, improve productivity, and generate higher-value goods and services. Investing in technology and creative industries adds diversity and drives future growth. Countries that prioritize the development of sectors like information technology, biotechnology, and renewable energy can become global leaders in innovation, attracting foreign investment, creating skilled jobs, and building a knowledge-based economy.

Furthermore, industrial diversification promotes inclusive economic development by creating employment opportunities across various sectors. Expanding into manufacturing, services, and creative industries allows for a more diverse workforce with various skills, reducing inequality by providing jobs in both high- and low-skilled sectors. It also encourages the growth of small and medium-sized enterprises (SMEs), which are critical drivers of economic development, entrepreneurship, and local innovation.

Environmental sustainability is another important benefit of diversification, particularly when economies move away from resource-intensive industries like mining or fossil fuel extraction and move toward more sustainable sectors. By investing in green technologies, renewable energy, and circular economies, countries can protect their environment and align their economies with global sustainability goals. This transition supports long-term ecological health and reduces the risk of environmental degradation accompanying dependence on extractive industries.

Industrial diversification is critical for building resilient, innovative, and inclusive economies. By shifting economic focus to a broader range of sectors—including manufacturing, services, technology, and creative industries—countries can reduce their dependency on single industries, drive innovation, create more job opportunities, and promote sustainable growth.

As global economies become increasingly interconnected and volatile, diversification serves as a buffer against external shocks and paves the way for a more stable, prosperous, and sustainable future.

9.1.2 Innovation and technology: Catalysts for economic growth and competitiveness

Innovation and technology drive productivity, competitiveness, and sustainable economic growth. Embracing new technologies and fostering a culture of innovation allows businesses and economies to adapt to an increasingly complex and interconnected global environment. By investing in research and development (R&D), adopting digital tools, and encouraging entrepreneurship, countries and companies can enhance their competitiveness, create new industries, and increase overall economic resilience.

One of the primary benefits of innovation is its ability to boost productivity. Technological advancements like automation, artificial intelligence (AI), and data analytics enable businesses to streamline processes, reduce costs, and improve efficiency. For example, automation in manufacturing can increase production speed and accuracy while minimizing human error. Similarly, AI and machine learning can optimize supply chains, improve decision-making, and lead to better outcomes. These productivity improvements contribute directly to economic growth by allowing businesses to produce more with fewer resources, ultimately enhancing competitiveness in the global market.

Investing in research and development is critical for fostering innovation. R&D activities generate new knowledge, products, and technologies that can transform industries and create new markets. Governments and private enterprises prioritizing R&D stimulate technological progress and promote long-term economic growth by laying the foundation for future innovations. Additionally, R&D investment often leads to the creation of intellectual property, such as patents and copyrights, which can provide competitive advantages and drive profitability. Countries that allocate significant resources to R&D, like South Korea, Japan, and Germany, have seen tremendous gains in innovation and global market leadership.

Adopting digital tools is another essential component of driving innovation. In today's digital age, technologies like cloud computing, the Internet of Things (IoT), and blockchain are revolutionizing industries and enabling businesses to operate more efficiently. Cloud-based platforms allow for scalable and cost-effective solutions, while IoT technologies connect devices to improve data collection and operational efficiency. With its decentralized and transparent approach, blockchain transforms finance, supply chain management, and healthcare industries. By integrating these digital tools, businesses can enhance agility, improve customer experiences, and open up new avenues for growth.

Entrepreneurship plays a vital role in harnessing innovation and technology. Entrepreneurs are often at the forefront of developing and commercializing disruptive technologies. By fostering an entrepreneurial ecosystem that encourages creativity, risk-taking, and innovation, countries can stimulate the growth of startups and small businesses, critical drivers of economic diversification and job creation. Policies that support access to capital, mentorship, and business networks can help entrepreneurs bring innovative solutions to the market. Government programs and public-private partnerships can also provide resources for early-stage startups, further catalyzing innovation.

Embracing innovation and technology has broader societal and environmental implications. Technological advancements in clean energy, for example, have the potential to reduce carbon emissions and address global climate change significantly. Innovations in healthcare, such as telemedicine and precision medicine, can improve the quality of life and extend life expectancy. By leveraging technology to solve pressing social and environmental issues, businesses and governments can promote sustainable development while maintaining economic growth.

Innovation and technology drive productivity, enhance competitiveness, and create sustainable economic growth. By investing in R&D, adopting digital tools, and encouraging entrepreneurship, economies can build a more resilient, diversified, and prosperous future. As industries continue to evolve, those that embrace technological change and foster a culture of innovation will be best positioned to thrive in the global marketplace.

9.1.3 Human capital development: A pillar of economic progress

Human capital development refers to improving a population's skills, knowledge, and well-being to enhance its productivity and ability to adapt to changing economic conditions. Investing in education, training, and healthcare is essential for cultivating a capable workforce that will drive economic growth, foster innovation, and sustain social progress. Countries and organizations that prioritize human capital development are better positioned to compete in the global economy, adapt to technological advances, and address emerging challenges.

A well-educated workforce is critical for sustained economic success. Education equips individuals with the skills, knowledge, and critical thinking abilities needed to participate effectively in the workforce. Whether basic literacy and numeracy or advanced technical and professional skills, education empowers people to contribute meaningfully to their industries and communities. Countries investing in education, such as Finland and South Korea, have seen remarkable economic output and innovation improvements. Moreover, higher levels of education are closely linked to increased individual income, improved job security, and greater social mobility, contributing to a more equitable society.

Training and skills development are equally important, especially in a rapidly changing global economy. As new technologies reshape industries and create new job opportunities, the demand for workers with specialized skills continues to grow. Investment in vocational training, technical education, and continuous learning programs ensures workers stay relevant and competitive. This benefits individuals by improving their employability and strengthens businesses and economies by ensuring access to a highly skilled workforce. For instance, Germany's dual education system, which combines apprenticeships with classroom learning, has been instrumental in maintaining the country's competitive edge in industries like manufacturing and engineering.

Healthcare is another fundamental aspect of human capital development. A healthy population is more productive, as individuals with good physical and mental health can work efficiently and contribute to the economy without illness or disability. Investments in healthcare improve life expectancy, reduce absenteeism, and enhance the overall well-being of the workforce. For example, countries with robust healthcare systems, such as Norway and Japan, often experience lower healthcare costs in the long term due to preventive care and improved public health. Moreover, access to healthcare services ensures that workers are physically and mentally fit to adapt to changing economic conditions, enhancing resilience in the face of pandemics or economic downturns.

In addition to increasing productivity, human capital development fosters adaptability in the workforce. As economic conditions evolve due to globalization, technological advancements, and shifts in market demand, workers must be able to transition to new roles and industries. Education and training programs emphasizing lifelong learning and upskilling allow individuals to stay competitive in the job market, ensuring that economies remain resilient to change. For example, as automation and artificial intelligence continue transforming industries, workers with access to reskilling programs are better equipped to transition from traditional jobs to emerging fields like data science, renewable energy, and digital services.

Human capital development reduces inequality and fosters social cohesion. Access to quality education, training, and healthcare can break cycles of poverty and enable individuals from all socioeconomic backgrounds to improve their livelihoods. By leveling the playing field and providing equal opportunities, societies can reduce disparities and create more inclusive economic growth. This leads to a more engaged and productive workforce, as individuals are empowered to contribute to society regardless of their background.

Human capital development is a cornerstone of economic and social progress. Governments and organizations can build a well-prepared, adaptable, and resilient workforce by investing in education, training, and healthcare. This drives productivity and innovation, fosters social equity, improves quality of life, and ensures long-term economic sustainability. In a rapidly changing world, economies prioritizing human capital will be better equipped to thrive in the face of new challenges and opportunities.

9.1.4 Sustainable practices: Balancing economic growth and environmental stewardship

Sustainable practices involve integrating environmental sustainability into economic activities to ensure that economic growth does not come at the cost of environmental degradation. In a world facing climate change, resource depletion, and biodiversity loss, adopting sustainable practices has become necessary for businesses, governments, and individuals. These practices include promoting renewable energy, enhancing resource efficiency, and investing in green technologies, all of which help minimize the environmental impact of economic activities while fostering long-term economic growth.

One of the most critical components of sustainable practices is the transition to renewable energy. Fossil fuel consumption has been a major driver of global warming and environmental degradation, but renewable energy sources like solar, wind, and hydropower offer cleaner alternatives. By investing in and adopting these renewable technologies, economies can reduce their reliance on carbon-intensive energy sources, significantly lowering greenhouse gas emissions. Countries like Denmark and Costa Rica have successfully implemented policies to harness renewable energy, achieving considerable reductions in their carbon footprints and creating jobs in green industries. This shift benefits the environment, strengthens energy security, and stimulates economic diversification.

Resource efficiency is another essential aspect of sustainable practices. It uses natural resources—such as water, minerals, and energy—more efficiently throughout production and consumption. Resource efficiency helps businesses reduce costs and waste while preserving ecosystems for future generations. For example, circular economy principles focus on minimizing waste by recycling materials, designing products for longevity, and reducing resource use through innovation. Companies that adopt circular economy models can improve their bottom line by cutting waste and conserving resources, contributing to a more sustainable and competitive business environment. Industries that optimize resource use also enhance their resilience, as they are less vulnerable to resource scarcity or price volatility.

Green technologies are crucial in advancing sustainable practices by enabling cleaner, more efficient production processes and reducing environmental impacts. Innovations such as electric vehicles, energy-efficient buildings, and sustainable agriculture practices transform industries, helping reduce carbon emissions, conserve water, and protect biodiversity. Green technologies are also crucial to addressing global challenges such as climate change, pollution, and habitat destruction. For example, precision agriculture technologies allow farmers to use fewer inputs like water and pesticides while increasing yields, making agriculture more sustainable. By investing in research and development of green technologies, businesses, and governments can create new opportunities for growth while contributing to environmental sustainability.

The integration of sustainable practices is not only an environmental necessity but also an economic opportunity. As global demand for sustainable products and services increases, businesses that embrace green technologies and resource-efficient practices gain a competitive edge. Consumers are becoming more conscious of the environmental impact of their purchases, and companies that prioritize sustainability are increasingly rewarded in the marketplace. Additionally, regulatory pressures, such as carbon pricing and stricter environmental standards, are pushing industries to adopt sustainable practices or risk facing higher costs and reduced profitability. Those who adapt early benefit from the growing green economy, while those who resist may face economic disadvantages.

Sustainable practices can contribute to job creation and economic resilience. The green economy, driven by investments in renewable energy, green infrastructure, and sustainable industries, generates millions of jobs worldwide. The International Labor Organization (ILO) estimates that transitioning to a greener economy could create 24 million new jobs globally by 2030. These jobs span sectors like clean energy, sustainable agriculture, and eco-friendly construction, providing new employment opportunities in industries prioritizing environmental sustainability. This shift to a green economy also helps reduce dependency on finite natural resources, building more resilient economies better equipped to cope with resource constraints and environmental challenges.

Sustainable practices are essential for aligning economic activities with environmental stewardship. Economies can minimize their environmental impact by promoting renewable energy, enhancing resource efficiency, and investing in green technologies while fostering long-term growth and resilience. Integrating sustainability into economic activities helps mitigate environmental risks and creates opportunities for innovation, job creation, and competitiveness in the emerging green economy. As global environmental challenges intensify, sustainable practices will be crucial in building a more prosperous and sustainable future for both people and the planet.

9.1.5 Institutional reforms: Building a strong foundation for economic growth

Institutional reforms are critical in creating a conducive environment for sustainable economic growth. Strong institutions and good governance are essential for ensuring economic activities are conducted fairly, transparently, and accountable. Countries can foster a more stable, predictable, and competitive economic climate by improving legal frameworks, reducing corruption, and strengthening public administration. These reforms enhance economic performance, promote social cohesion, reduce inequalities, and increase public trust in government institutions.

A well-functioning legal framework is the backbone of any healthy economy. Laws governing property rights, contracts, business operations, and dispute resolution provide the structure that supports economic activities. Without clear and enforceable legal frameworks, businesses face uncertainty, which can deter investment and hinder economic growth. For instance, secure property rights are crucial for encouraging entrepreneurship, as individuals and companies are more likely to invest in land, buildings, and equipment if they are confident that the law will protect their assets. Effective contract enforcement is similarly vital, as it ensures that businesses can confidently engage in transactions, knowing that the legal system will uphold agreements. Countries implementing legal reforms to protect property rights and strengthen contract enforcement often see improvements in business activity, investment, and overall economic development.

Reducing corruption is another crucial aspect of institutional reform. Corruption undermines public trust in government and distorts economic activities by diverting resources away from productive uses and into the hands of a few. It creates inefficiencies, raises business costs, and discourages foreign and domestic investment. Corruption also exacerbates inequality, as those with power and connections often benefit at the expense of the broader population. Institutional reforms aimed at reducing corruption—such as increasing transparency, strengthening anti-corruption agencies, and ensuring the accountability of public officials—can lead to more equitable economic outcomes and foster a more competitive business environment. Countries that have successfully tackled corruption, such as Singapore and Estonia, have seen significant improvements in governance, economic performance, and investor confidence.

Effective public administration is also essential for creating a conducive environment for economic activities. Public institutions that are well-managed, transparent, and accountable ensure the efficient delivery of public services, such as education, healthcare, and infrastructure, which are critical for economic growth. Strong public administration allows governments to implement policies that promote economic development, regulate markets effectively, and provide a social safety net for vulnerable populations. When public institutions are weak or inefficient, the delivery of these services suffers, which can hinder economic performance and exacerbate inequality. Institutional reforms aimed at improving public administration—such as modernizing bureaucratic processes, adopting digital governance tools, and enhancing the professionalism of civil servants—can lead to more efficient government operations and a more favorable environment for businesses and citizens alike.

Institutional reforms also play a crucial role in creating a predictable regulatory environment for attracting investment and promoting long-term economic stability. Domestic and foreign investors are more likely to commit resources when they have confidence in the consistency and fairness of a country's legal and regulatory frameworks. Unpredictable regulations or sudden policy changes can create uncertainty, leading businesses to delay or abandon investment plans. By reforming regulatory institutions to be more transparent, consistent, and responsive to the needs of businesses, governments can foster a climate of stability that encourages innovation, entrepreneurship, and growth. For example, countries that have streamlined their regulatory processes, such as South Korea and New Zealand, have seen a surge in business activity and foreign investment, boosting overall economic competitiveness.

Institutional reforms also promote social cohesion by reducing inequality and enhancing public trust in government. When legal frameworks are fair, corruption is minimized, and public institutions are efficient, citizens are more likely to feel that they are being treated equitably, which fosters social stability and reduces conflict. This is particularly important in diverse societies where economic and social inequalities can lead to tensions. By strengthening institutions and governance, countries can create a more inclusive environment where all citizens can access opportunities and contribute to economic growth. This, in turn, promotes long-term stability and reduces the likelihood of social unrest or political instability, which can have negative economic consequences.

Institutional reforms are vital for creating a strong foundation for economic activities. Governments can foster a more transparent, predictable, and competitive economic environment by strengthening legal frameworks, reducing corruption, and improving public administration. These reforms enhance economic performance, promote social stability, reduce inequalities, and build public trust in institutions. Countries prioritizing institutional reforms in an increasingly globalized world will be better positioned to achieve sustained economic growth and long-term prosperity.

9.1.6 Inclusive growth: Ensuring economic benefits for all

Inclusive growth is a fundamental aspect of sustainable economic transformation, aiming to ensure that development benefits are shared equitably across all segments of society. Rapid economic growth has often led to widening inequality in many countries, excluding large portions of the population from its advantages. Inclusive growth seeks to address these disparities by reducing income inequality, providing social protection, and creating opportunities for marginalized groups. It is an approach that promotes economic efficiency, social equity, and cohesion, fostering long-term stability and prosperity for all.

One of the primary goals of inclusive growth is to address income inequality. In many economies, the gap between the rich and the poor has been widening, with wealth concentrating in the hands of a few, while large portions of the population remain in poverty. Inclusive growth policies aim to reduce this inequality by ensuring that economic gains are distributed more evenly. This can be achieved through progressive taxation, wage policies that support a living wage, and targeted public investments in sectors that create jobs for low- and middle-income earners. Countries like Sweden and Denmark have implemented policies prioritizing income redistribution and social welfare, resulting in lower inequality levels and stronger social safety nets. By narrowing the income gap, inclusive growth improves living standards for the poorest members of society. It strengthens overall economic demand as more people have the purchasing power to participate in the economy.

Providing social protection is another critical component of inclusive growth. Social protection systems—healthcare, education, pensions, and unemployment benefits—are essential for reducing poverty, protecting vulnerable populations, and ensuring everyone can participate in the economy. These systems act as buffers during economic hardship, helping individuals and families maintain a basic standard of living even in the face of unemployment, illness, or other life challenges. Inclusive growth policies prioritize expanding social protection programs to ensure no one is left behind. For instance, Brazil's Bolsa Família program has provided financial assistance to millions of low-income families, significantly reducing poverty and improving access to education and healthcare. By investing in social protection, governments can foster more resilient populations capable of adapting to economic shifts while maintaining their well-being.

Creating opportunities for marginalized groups is also essential for achieving inclusive growth. Many populations—such as women, ethnic minorities, people with disabilities, and rural communities—are often excluded from the formal economy due to systemic barriers such as discrimination, lack of education, or limited access to resources. Inclusive growth policies aim to break down these barriers by creating an enabling environment where marginalized groups can participate fully in economic activities. This can involve improving access to education and vocational training, providing microfinance and entrepreneurial support, and enforcing anti-discrimination laws in the workplace. Countries like Rwanda have made significant strides in promoting gender equality by ensuring women can access education, healthcare, and leadership positions in government and business. By empowering marginalized groups, inclusive growth policies create a more diverse and productive workforce, driving innovation and fostering a more dynamic economy.

Inclusive growth also emphasizes the importance of job creation and decent work for all. Economic transformation often leads to shifts in the labor market, with some sectors shrinking while others expand. Inclusive growth policies aim to ensure workers are not left behind by these changes, focusing on skills development, retraining programs, and job placement services. For example, equipping workers with the skills needed to succeed in these industries is crucial as economies transition toward more knowledge-based and technology-driven sectors. Countries like Singapore have invested heavily in upskilling programs that allow workers to transition into high-demand fields such as technology and healthcare, ensuring that broad-based employment opportunities accompany economic growth. By prioritizing job creation and skills development, inclusive growth ensures that all members of society can benefit from economic transformation.

Inclusive growth policies recognize the importance of geographic inclusion. Economic growth is often concentrated in urban centers, leaving rural areas underserved and underdeveloped. Inclusive growth aims to bridge this urban-rural divide by investing in rural infrastructure, education, and healthcare and supporting smallholder farmers and rural entrepreneurs. By ensuring that rural populations have access to the same opportunities as urban dwellers, inclusive growth promotes balanced regional development and reduces migration pressures on cities. Countries like China have invested heavily in rural development programs, lifting millions out of poverty and bringing prosperity to previously left behind regions.

Inclusive growth is essential for ensuring that the benefits of economic transformation are shared equitably across society. Inclusive growth promotes economic efficiency and social equity by addressing income inequality, providing social protection, and creating opportunities for marginalized groups. It fosters a more resilient, innovative, and cohesive society where all members can contribute to and benefit from economic progress. As economies evolve in the face of globalization, technological advancements, and climate change, inclusive growth will remain a vital strategy for building a fairer and more sustainable future.

9.2 The Process of Economic Transformation

Economic transformation involves shifting the structure of an economy to promote sustainable growth, increased productivity, and higher living standards. It typically includes diversification from traditional sectors like agriculture or extraction to a broader range of industries, such as manufacturing, technology, and services. Key components include investing in innovation, technology, and human capital development, promoting sustainable practices, and implementing institutional reforms to improve governance. By fostering inclusive growth and ensuring equitable distribution of benefits, economic transformation aims to enhance resilience, reduce poverty, and create long-term economic opportunities for all segments of society.

9.2.1 Policy formulation and implementation: Guiding economic transformation

Policy formulation and implementation are essential processes through which governments guide and shape economic transformation. Governments create a framework that facilitates diversification, innovation, and sustainability by setting clear strategic plans and targets. The role of government extends beyond mere oversight; it involves actively crafting policies that promote growth, protect the environment, and enhance social equity. This process requires collaboration across various sectors, including public, private, and civil society, to ensure that policies are comprehensive and effective in achieving desired outcomes.

One of the primary functions of policy formulation is to establish strategic plans that outline the government's vision for economic transformation. These plans often identify key sectors for diversification and innovation, enabling countries to shift away from over-reliance on traditional industries. For instance, a government may recognize the need to transition from a predominantly agricultural economy to one that includes technology and manufacturing. By setting clear targets for each sector, governments can create measurable objectives that guide investment decisions and resource allocation. This strategic approach not only aligns public and private sector efforts but also communicates to stakeholders the priorities and direction of economic policy. Countries that have successfully implemented such strategic plans, like South Korea, which focuses on technology and export-driven growth, have seen significant economic advancements.

Incentivizing sustainable practices is another critical aspect of policy formulation. Governments can encourage businesses to adopt environmentally friendly practices through tax incentives, grants, and subsidies. For instance, offering tax breaks for companies that invest in renewable energy technologies or implement energy-efficient processes can motivate firms to prioritize sustainability in their operations. Additionally, grants for research and development (R&D) can spur innovation by providing the necessary funding for new ideas and technologies. By aligning financial incentives with sustainability goals, governments can create an ecosystem where businesses are motivated to contribute to environmental and social objectives while driving economic growth.

Educational reforms are equally vital in the context of economic transformation. As markets evolve and new industries emerge, the workforce must possess the skills to meet changing demands. Policymakers must prioritize educational programs that align with market needs, fostering a culture of lifelong learning and adaptability. This can involve enhancing vocational training, integrating technology into curricula, and promoting STEM (science, technology, engineering, and mathematics) education. Governments ensure their populations can thrive in a dynamic economy by investing in human capital. Countries like Germany, with its dual education system that combines classroom learning with practical training, serve as exemplary models of how educational reforms can support economic transformation.

Effective policy implementation is crucial for the success of formulated strategies. It is not enough to design policies; they must be effectively executed and monitored to achieve their intended outcomes. Governments should establish clear mechanisms for tracking progress and assessing the impact of policies on economic transformation. This may involve setting up regulatory bodies to oversee compliance, collecting data on performance indicators, and engaging with stakeholders to gather feedback. Transparent monitoring processes help identify challenges and enable governments to make necessary policy adjustments over time. Countries emphasizing robust implementation frameworks, like Singapore, have demonstrated how effective governance can drive successful economic transformation.

Collaboration and stakeholder engagement are also essential in policy formulation and implementation. Governments should actively involve businesses, civil society organizations, and other stakeholders in developing policies that affect their interests. This participatory approach fosters buy-in from various sectors and enhances the likelihood of successful implementation. For example, public-private partnerships can facilitate investment in infrastructure and technology, enabling more effective resource utilization and innovation. Engaging stakeholders in policy-making also ensures that diverse perspectives are considered, leading to more inclusive and equitable outcomes.

Policymakers must remain adaptable in evolving global economic conditions. As new challenges arise—such as climate change, technological disruption, and geopolitical shifts—governments must be prepared to adjust their strategies and policies accordingly. This requires a forward-thinking mindset that embraces innovation and flexibility. Countries that have successfully navigated economic transformation, such as Finland, focusing on education and technology, demonstrate the importance of agility in policy formulation and implementation. Governments can foster resilient economies that thrive in a rapidly changing world by continuously reassessing priorities and incorporating lessons learned.

Policy formulation and implementation are crucial for guiding economic transformation. Governments can create an environment conducive to innovation, diversification, and inclusive growth through strategic planning, incentives for sustainable practices, educational reforms, and effective governance. By actively engaging stakeholders and remaining adaptable to change, policymakers can ensure that the benefits of economic transformation are widely shared and contribute to long-term prosperity for all segments of society.

9.2.2 Investment in infrastructure: A cornerstone of economic transformation

Investment in infrastructure plays a pivotal role in driving economic transformation by providing the essential foundation for growth and development. Infrastructure encompasses many components, including physical assets such as transportation networks, energy systems, and telecommunications and soft infrastructure like education and healthcare systems. A robust infrastructure framework enhances productivity, facilitates trade, and improves the quality of life for citizens, ultimately laying the groundwork for sustainable economic progress.

Physical infrastructure is vital for ensuring the efficient movement of goods and services. Well-developed transportation networks, including roads, railways, ports, and airports, reduce travel time and costs, making it easier for businesses to access markets and supply chains. For example, countries with extensive rail networks, such as Germany, benefit from increased freight capacity and reduced transportation costs, promoting trade and economic connectivity. Similarly, investment in ports and logistics hubs can position a country as a regional trade center, attracting foreign investment and enhancing competitiveness. By improving physical infrastructure, governments create an environment conducive to business growth and economic activity, driving overall transformation.

Energy infrastructure is another critical component of economic transformation. Reliable and sustainable energy sources are essential for powering industries, homes, and public services. Investing in renewable energy infrastructure—such as solar, wind, and hydroelectric power—can help countries transition away from fossil fuels and reduce their carbon footprint. Moreover, access to reliable energy sources enables businesses to operate more efficiently, reducing production costs and enhancing competitiveness. For instance, countries like Denmark have made significant strides in renewable energy investments, leading to energy independence and sustainability. Governments can foster economic growth by prioritizing energy infrastructure while addressing environmental concerns.

Telecommunications infrastructure is equally important in today's digital economy. High-speed internet access and mobile connectivity enable communication, information exchange, and innovation. Investment in telecommunications infrastructure enhances digital inclusion, allowing businesses and individuals in rural or underserved areas to access online services, e-commerce, and educational resources. Countries prioritizing digital infrastructure, like South Korea, have seen significant economic benefits from increased connectivity and technological adoption. As digital economies expand, investing in telecommunications will foster innovation and ensure that all citizens can participate in the digital landscape.

Soft infrastructure, such as education and healthcare systems, is essential for fostering human capital development and improving the overall quality of life. A well-educated workforce is critical for driving innovation and adapting to changing economic conditions. Governments must invest in quality education systems that equip individuals with the skills and knowledge needed to thrive in a dynamic economy. This can involve expanding access to vocational training, higher education, and lifelong learning programs. Countries prioritizing education, like Finland, consistently rank high in global competitiveness and innovation, demonstrating the importance of investing in human capital.

Healthcare infrastructure also plays a vital role in economic transformation. A healthy population is more productive and capable of contributing effectively to the economy. Investments in healthcare systems ensure access to essential services, preventive care, and health education, ultimately improving public health outcomes. Moreover, robust healthcare infrastructure enhances resilience against economic shocks like pandemics or natural disasters. Countries with strong healthcare systems, like Switzerland, are better equipped to manage health crises and maintain economic stability. Governments promote public well-being and contribute to economic growth and stability by investing in healthcare infrastructure.

The quality of infrastructure directly impacts the business environment and investment climate. Inadequate infrastructure can deter foreign investment and limit economic opportunities. Investors seek locations with reliable transportation, energy, and communications systems, as these factors influence operational efficiency and profitability. By prioritizing infrastructure development, governments can attract domestic and foreign investment, creating a virtuous cycle of economic growth. Countries like Singapore have successfully leveraged their strategic investments in infrastructure to position themselves as global business hubs, demonstrating how infrastructure can catalyze economic transformation.

Investment in infrastructure is a cornerstone of economic transformation. By enhancing physical and soft infrastructure, governments create an environment that fosters productivity, innovation, and improved quality of life. From transportation and energy to education and healthcare, robust infrastructure is essential for driving sustainable growth and development. By prioritizing infrastructure investments and leveraging public-private partnerships, governments can position their economies for success in an increasingly competitive global landscape, ensuring that all citizens benefit from the fruits of economic transformation.

9.2.3 Public-private partnerships: Catalysts for economic transformation

Public-private partnerships (PPPs) represent a strategic collaboration between government entities and private sector organizations to achieve shared goals in infrastructure development, innovation, and sustainable practices. As economies face growing challenges such as urbanization, climate change, and technological advancement, the role of PPPs in facilitating economic transformation becomes increasingly vital. By leveraging the private sector's expertise, capital, and efficiency, governments can address pressing needs and drive sustainable growth, creating a more resilient and competitive economy.

One of the primary benefits of PPPs is their ability to mobilize private sector investment in infrastructure development. Governments often face budget constraints and limited resources for large-scale projects, which can delay critical infrastructure improvements. Through PPPs, private companies can contribute capital and resources, reducing the financial burden on public budgets. This collaboration accelerates infrastructure projects essential for economic growth, such as transportation networks, energy facilities, and social infrastructure. For example, developing highways and expressways through PPPs in India has significantly improved connectivity, reduced travel time, and stimulated economic activity in various regions.

PPPs have emerged as a powerful tool for advancing sustainable practices in renewable energy. As the world grapples with the urgent need to transition to clean energy sources, governments can partner with private companies to develop renewable energy projects such as solar farms, wind power installations, and hydroelectric facilities. These partnerships not only harness private sector expertise in technology and project management but also facilitate the sharing of risks and rewards associated with investment in clean energy. For instance, in South Africa, the Renewable Energy Independent Power Producer Procurement Program (REIPPPP) has attracted private investment in renewable energy projects, leading to increased energy capacity and reduced carbon emissions. Such initiatives exemplify how PPPs can drive the shift toward a more sustainable energy landscape while fostering economic development.

PPPs are also critical in promoting innovation, particularly in emerging sectors like smart cities and digital infrastructure. As urbanization accelerates, cities increasingly seek innovative solutions to improve service delivery, enhance quality of life, and address environmental challenges. By collaborating with private technology firms, governments can implement smart city initiatives that utilize data analytics, Internet of Things (IoT) devices, and digital platforms to optimize urban services such as transportation, waste management, and public safety. For example, Barcelona's smart city initiatives leverage data-driven technologies to improve urban mobility and resource management, enhancing efficiency and sustainability. Such collaborations enable governments to tap into the innovative capabilities of the private sector, driving economic transformation through advanced technological solutions.

PPPs can foster sustainable development by integrating environmental considerations into infrastructure projects. Governments can work with private partners to prioritize sustainability and resilience when designing and implementing infrastructure initiatives. This may involve incorporating green building practices, utilizing renewable materials, and minimizing environmental impact during construction and operation. For instance, developing eco-friendly public transportation systems through PPPs can reduce greenhouse gas emissions and promote sustainable mobility. By aligning private sector incentives with public policy goals, PPPs can drive sustainable practices across various sectors, contributing to broader environmental objectives.

However, PPPs' success hinges on effective governance, transparency, and stakeholder engagement. Clear frameworks and guidelines are essential for establishing the roles and responsibilities of each partner, ensuring accountability and alignment with public interests. Fostering open communication and collaboration between public and private entities can also build trust and facilitate knowledge sharing. Governments must also engage with local communities and stakeholders to gather input and address concerns related to infrastructure projects, ensuring that developments meet the needs of those affected. For instance, in Canada, the Toronto Waterfront Revitalization Corporation actively involves community members in the planning and implementation of waterfront projects through consultations and participatory processes, leading to more inclusive and successful outcomes.

Challenges also exist within PPPs, particularly concerning risk-sharing and financial arrangements. Governments must carefully consider the distribution of risks and rewards between public and private partners to avoid potential conflicts and ensure long-term sustainability. Furthermore, the financial viability of projects must be assessed to attract private investment, requiring thorough feasibility studies and transparent procurement processes. Countries that have effectively navigated these challenges, such as Australia and the United Kingdom, provide valuable lessons on best practices for structuring and implementing successful PPPs.

Public-private partnerships are essential for fostering economic transformation by leveraging the strengths of both the public and private sectors. Governments can mobilize private investment, drive innovation, and promote sustainable practices across various sectors through collaboration. By developing robust frameworks, engaging stakeholders, and addressing potential challenges, PPPs can effectively contribute to infrastructure development, renewable energy projects, and smart city initiatives, ultimately enhancing the quality of life for citizens and ensuring a more sustainable and resilient economy. As the global economy continues to evolve, the importance of effective partnerships between public and private entities will only grow, providing a pathway for sustainable economic progress.

9.2.4 Support for small and medium enterprises (SMEs)

Small and medium enterprises (SMEs) are often regarded as the backbone of economies, playing a crucial role in driving economic transformation, innovation, and job creation. These businesses, characterized by their smaller scale compared to large corporations, contribute significantly to employment and gross domestic product (GDP) in many countries. Supporting SMEs is essential for fostering a vibrant economy and ensuring sustainable development, as they provide diverse products and services while contributing to local communities. Governments, financial institutions, and other stakeholders must prioritize support for SMEs to unlock their full potential and drive economic growth.

Access to finance is one of the most pressing challenges faced by SMEs. Traditional financial institutions often perceive these businesses as high-risk investments due to their limited collateral, cash flow volatility, and lack of credit history. As a result, SMEs frequently struggle to secure the funding needed for expansion, innovation, and operational improvements. To address this issue, governments and financial institutions can implement targeted programs that provide loans, grants, and microfinancing options tailored to the unique needs of SMEs. For instance, establishing government-backed loan guarantees can reduce the risk for lenders, encouraging them to extend credit to small businesses. Additionally, innovative financing mechanisms, such as crowdfunding platforms, can provide alternative sources of capital for SMEs, enabling them to access funds more easily.

SMEs require access to business development services to enhance their competitiveness and growth prospects. These services may include mentorship, training, and technical assistance to help business owners develop the skills and knowledge necessary for effective management and operations. Governments and non-profit organizations can play a vital role in facilitating capacity-building initiatives, such as workshops and training programs, to equip SME owners with essential business skills, marketing strategies, and financial management techniques. By investing in the human capital of SMEs, stakeholders can foster a more resilient and adaptable workforce, ultimately driving innovation and productivity across the economy.

Market access is another critical factor influencing the success of SMEs. Many small businesses struggle to penetrate larger markets or compete with established corporations due to lacking resources and visibility. To support SMEs in overcoming these barriers, governments can implement policies that promote fair competition and create opportunities for small businesses to participate in public procurement processes. For example, setting aside a portion of government contracts specifically for SMEs can enhance their access to lucrative opportunities, enabling them to grow and create jobs. Additionally, fostering partnerships between SMEs and larger corporations can facilitate knowledge sharing, technology transfer, and market entry, further enhancing the competitiveness of small businesses.

Innovation is a key driver of economic transformation, and SMEs are often at the forefront of this process. Their agility and capacity for creative problem-solving enable them to develop innovative products and services that can address emerging market needs. Governments and financial institutions should establish programs encouraging research and development (R&D) within SMEs to nurture this innovation. This can include offering tax incentives for R&D expenditures, creating innovation grants, and facilitating collaboration between SMEs and research institutions. By fostering a culture of innovation, stakeholders can help SMEs remain competitive in an ever-changing global market and contribute to broader economic transformation.

The importance of supporting SMEs extends beyond economic growth; it also encompasses social benefits. SMEs often play a vital role in job creation, particularly in local communities where large corporations may not establish a presence. By providing employment opportunities, SMEs contribute to poverty alleviation and social mobility, helping to build stronger communities. Furthermore, supporting women-owned and minority-owned SMEs can promote inclusivity and diversity within the economy, addressing historical inequalities and empowering underrepresented groups. Governments can implement targeted initiatives to support these businesses, such as providing tailored financial assistance and capacity-building programs.

Challenges also exist in supporting SMEs, including the need for effective monitoring and evaluation of assistance programs. Policymakers must ensure that resources are allocated efficiently and that support initiatives are tailored to the specific needs of SMEs. Engaging with SME owners and stakeholders in developing and implementing support programs can help ensure that initiatives are responsive and effective. Additionally, fostering collaboration between various stakeholders, including government agencies, financial institutions, and business associations, can enhance the effectiveness of support efforts and create a more conducive environment for SME growth.

Support for small and medium enterprises drives economic transformation, fosters innovation, and creates job opportunities. By enhancing access to finance, providing business development services, improving market access, and promoting innovation, stakeholders can empower SMEs to thrive and contribute to sustainable economic growth. As economies continue to evolve, prioritizing support for SMEs will be essential to harnessing their potential and ensuring that the benefits of economic transformation are shared widely across society. By recognizing the critical role of SMEs in economic development, governments and financial institutions can create a more resilient and inclusive economy that supports all citizens.

9.2.5 Enhancing trade and investment: A pathway to economic transformation

Enhancing trade and investment is a pivotal strategy for stimulating economic transformation, especially in today's interconnected global economy. By opening markets, promoting trade, and attracting foreign direct investment (FDI), countries can create a conducive environment for growth, innovation, and competitiveness. Integrating local economies into global value chains allows nations to leverage their unique strengths, access new markets, and benefit from international best practices. Governments can play a crucial role in facilitating this process through targeted policies, trade agreements, and investments in infrastructure.

One of the primary means of enhancing trade is establishing trade agreements. Bilateral and multilateral agreements can reduce tariffs, eliminate trade barriers, and foster collaboration between countries. These agreements often create a more predictable and stable trade environment, encouraging businesses to explore international markets. For example, the North American Free Trade Agreement (NAFTA) facilitated trade between the United States, Canada, and Mexico, resulting in significant economic growth and increased competitiveness for participating nations. By reducing trade barriers, countries can attract foreign investors seeking access to these markets, creating a win-win scenario for all parties involved.

Foreign direct investment (FDI) is another critical economic transformation driver. FDI brings capital, technology, and expertise into local economies, fostering innovation and improving productivity. To attract FDI, governments can implement investment promotion strategies that showcase their market potential, skilled labor force, and favorable regulatory environment. This includes establishing dedicated investment promotion agencies to assist foreign investors in navigating the local landscape and addressing any concerns related to doing business in the country. Streamlining bureaucratic processes and reducing red tape can significantly improve business efficiency, making the country more appealing to potential investors.

Special Economic Zones (SEZs) are a practical example of how enhancing trade and investment can spur economic transformation. SEZs are designated areas within a country that operate under different economic regulations than the rest. These zones often offer tax incentives, streamlined customs procedures, and infrastructure support to attract foreign investment and promote industrial development. By creating an attractive environment for businesses, SEZs can serve as engines of growth, generating jobs and facilitating technology transfer. Countries such as China and India have successfully leveraged SEZs to stimulate economic activity and attract significant foreign investment, demonstrating the effectiveness of this approach.

Integrating local economies into global value chains can lead to sustainable economic transformation. Countries can access larger markets and increase competitiveness by participating in global supply chains. This integration often necessitates improvements in infrastructure, logistics, and skills development to meet international standards. For instance, countries that invest in improving transportation networks and digital connectivity can facilitate smoother trade flows and enhance their attractiveness to foreign investors. Additionally, fostering partnerships between local businesses and multinational corporations can create opportunities for knowledge sharing, technology transfer, and capacity building, ultimately driving innovation and growth.

Enhancing trade and investment is also closely linked to pursuing sustainable development goals. Trade can stimulate economic growth while ensuring that environmental and social considerations are considered. By promoting sustainable trade practices, governments can encourage businesses to adopt environmentally friendly production methods and adhere to social standards. For example, countries can implement regulations that require businesses to demonstrate responsible sourcing practices or commit to reducing carbon emissions. This not only attracts socially conscious investors but also enhances the long-term viability of local economies. However, challenges remain in enhancing trade and investment. Geopolitical tensions, trade wars, and protectionist policies can hinder the flow of goods and capital across borders. Additionally, ensuring that the benefits of trade and investment are distributed equitably within society is essential for maintaining social stability and support for open markets. Policymakers, with their crucial role in balancing the interests of various stakeholders, are empowered to ensure that trade agreements and investment strategies contribute to inclusive growth and social well-being.

Enhancing trade and investment is a fundamental pathway to economic transformation. Countries can attract foreign capital, foster industrial development, and integrate into global value chains through trade agreements, targeted investment promotion, and the establishment of special economic zones. By creating a favorable business environment and investing in infrastructure, nations can unlock their economic potential, drive innovation, and enhance competitiveness. As the global economy continues to evolve, enhancing trade and investment will be crucial for achieving sustainable growth and ensuring that the benefits of economic transformation are shared widely across society.

9.2.6 Education and skill development

In an era characterized by rapid technological advancement and shifting labor market demands, reforming education systems to align with the needs of the modern economy has become increasingly crucial. Education and skill development serve as the foundation for a competitive workforce capable of adapting to the complexities of a globalized economy. By updating curricula, promoting STEM (science, technology, engineering, and mathematics) education, and emphasizing vocational training, nations can cultivate a workforce that meets current demands and is also equipped for the future.

A significant aspect of education reform is the modernization of curricula to ensure relevance in an ever-changing job market. Traditional educational models often lag behind industry needs, leaving graduates ill-prepared for employment. To address this gap, education systems must incorporate skills and knowledge that align with the demands of emerging industries. For example, integrating digital literacy and coding into the curriculum can prepare students for careers in technology while emphasizing critical thinking and problem-solving skills, which can enhance their employability across various sectors. Collaborative efforts between educational institutions and industry leaders can facilitate the development of curricula that reflect real-world applications and prepare students for successful careers.

Promoting STEM education is essential for equipping the workforce with the skills needed in a technology-driven economy. STEM fields are at the forefront of innovation and economic growth, driving advancements in various sectors, from healthcare to manufacturing. By prioritizing STEM education, countries can foster a culture of innovation and creativity, encouraging students to pursue careers in these critical areas. Initiatives such as after-school programs, mentorship opportunities, and partnerships with technology companies can inspire students to explore STEM disciplines and develop the skills required for high-demand jobs. Additionally, outreach programs aimed at underrepresented groups in STEM can help bridge the gender and diversity gaps, ensuring that all individuals have equal access to opportunities in these fields.

Vocational training and apprenticeships are vital in preparing individuals for skilled trades and technical occupations. As economies evolve, the demand for skilled construction, manufacturing, and healthcare workers continues to grow. By investing in vocational training programs, governments can provide individuals with hands-on experience and practical skills directly applicable to the workforce. Collaborating with industry partners to design training programs that align with labor market needs can enhance the effectiveness of vocational education. Furthermore, promoting the value of skilled trades can change societal perceptions, encouraging more individuals to pursue these rewarding career paths.

Lifelong learning initiatives are critical for ensuring the workforce remains adaptable and skilled throughout their careers. In a rapidly changing job market, the ability to upskill and reskill is essential for maintaining employability. Governments, educational institutions, and employers must collaborate to create accessible and flexible learning opportunities for individuals at all stages of their careers. This can include online courses, micro-credentials, and professional development programs that enable individuals to acquire new skills and knowledge relevant to their jobs. By fostering a culture of lifelong learning, societies can empower their workforce to navigate the complexities of the modern economy and seize emerging opportunities.

Furthermore, the integration of technology in education has the potential to enhance learning outcomes and accessibility. Digital tools can facilitate personalized learning experiences, enabling students to progress at their own pace and focus on areas where they need improvement. Online learning platforms can expand access to education for individuals in remote or underserved communities, breaking down geographical barriers. Additionally, incorporating technology into the classroom can prepare students for a tech-driven workforce, equipping them with the skills to thrive in various industries.

While the benefits of education and skill development are clear, implementing effective reforms remains challenging. Funding disparities, outdated infrastructure, and resistance to change can hinder progress in education systems. Policymakers must address these challenges by prioritizing investment in education, fostering collaboration between stakeholders, and creating supportive policies that encourage innovation in teaching and learning.

Education and skill development are essential components of economic transformation. By reforming education systems, promoting STEM education, emphasizing vocational training, and fostering lifelong learning, nations can cultivate a skilled and adaptable workforce capable of thriving in the modern economy. As the work landscape evolves, prioritizing education and skill development will ensure sustainable economic growth and empower individuals to realize their full potential.

9.2.7 Environmental sustainability

Incorporating environmental sustainability into economic transformation is essential for ensuring the long-term viability of both economies and ecosystems. By adopting sustainable practices, economies can grow in ways that protect natural resources, reduce environmental degradation, and address global challenges like climate change. This approach is not only crucial for the health of the planet but also for creating resilient, future-ready economies that can thrive in an increasingly resource-constrained world. Central to this transformation are green technologies, circular economy practices, and sustainable resource management policies, all of which enable a balanced relationship between economic growth and environmental stewardship.

Adopting green technologies is one of the most significant steps toward achieving environmental sustainability. These technologies help reduce the environmental impact of economic activities, enhance energy efficiency, and minimize waste. Renewable energy technologies, such as solar, wind, and hydropower, are at the forefront of this movement, providing cleaner alternatives to fossil fuels and significantly reducing carbon emissions. Transitioning from coal, oil, and natural gas to renewable energy sources not only addresses the climate crisis but also reduces dependence on finite, polluting resources, thereby ensuring energy security for the future. Many countries are now investing in green infrastructure projects that promote renewable energy use, such as solar farms and wind turbines, to power cities and industries sustainably.

In addition to renewable energy, the circular economy is another vital concept in environmental sustainability. The traditional linear economy model—take, make, dispose—leads to excessive waste and resource depletion. In contrast, the circular economy promotes reusing, recycling, and regenerating products and materials, reducing waste and conserving resources. This approach ensures that economic activities create minimal environmental impact by extending the product lifecycle and efficiently using raw materials. For example, industries can adopt circular practices such as designing products for easier disassembly and recycling, using biodegradable materials, and developing systems for recovering and reusing waste. The shift to a circular economy protects the environment, fosters innovation, creates new markets, and enhances competitiveness.

Sustainable resource management is another critical element of environmentally responsible economic transformation. Effective management of natural resources, such as water, forests, minerals, and land, ensures that economic activities do not exhaust or damage these resources for future generations. Governments play a crucial role by implementing policies and regulations promoting responsible resource extraction and use. For instance, sustainable forestry practices ensure that timber harvesting does not lead to deforestation and biodiversity loss. Similarly, adopting water conservation measures in agriculture and industry can protect freshwater resources from depletion and contamination. Such sustainable management policies are essential for balancing resource use and environmental health.

Sustainability policies also foster corporate responsibility and incentivize businesses to adopt greener practices. Governments can implement tax breaks or subsidies for companies that invest in green technologies, renewable energy, or sustainable production methods. Environmental regulations, such as carbon pricing or emissions trading systems, can encourage businesses to reduce their carbon footprint and invest in more sustainable operations. These policies are crucial for ensuring that businesses proactively address environmental issues while continuing to drive economic growth.

Furthermore, sustainable practices benefit the environment and create economic opportunities. The green economy has the potential to generate millions of new jobs in sectors like renewable energy, energy efficiency, waste management, and sustainable agriculture. Investing in these areas can stimulate job creation while contributing to a more sustainable and resilient economy. As more businesses and governments adopt green practices, the demand for environmentally friendly products and services will grow, creating new markets and innovation opportunities.

A prime example of sustainable economic transformation can be seen in countries that have successfully transitioned to renewable energy. Nations like Denmark and Germany have heavily invested in wind and solar energy, significantly reducing their reliance on fossil fuels while maintaining robust economic growth. These countries have cut their greenhouse gas emissions and positioned themselves as global leaders in green technology, creating jobs and boosting economies by exporting renewable energy solutions.

Environmental sustainability is integral to achieving a balanced, resilient, and future-oriented economic transformation. By adopting green technologies, promoting circular economy practices, and implementing sustainable resource management policies, economies can grow harmoniously with the planet's ecological limits. This approach addresses pressing global challenges like climate change and resource depletion and creates opportunities for innovation, job creation, and long-term economic stability. The shift toward sustainability is necessary for the planet's health and a strategic pathway for building a prosperous and equitable global economy.

9.2.8 Monitoring and evaluation: A strategic approach to economic transformation

Monitoring and evaluating policies and initiatives are critical to ensuring that economic transformation efforts remain on track and achieve their intended goals. Without robust monitoring and evaluation (M&E) frameworks, it is difficult to assess the effectiveness of policies, identify potential obstacles, and make the necessary course corrections to optimize outcomes. By systematically tracking progress and utilizing data-driven decision-making, governments and stakeholders can fine-tune their strategies and ensure that economic transformation efforts align with long-term objectives such as sustainability, inclusivity, and resilience.

The first step in effective M&E is setting clear, measurable goals for economic transformation initiatives. Whether these goals involve increasing GDP, creating jobs, reducing carbon emissions, or promoting industrial diversification, having specific targets allows policymakers to gauge progress over time. These targets must be accompanied by key performance indicators (KPIs) that track various aspects of economic development, such as productivity levels, investment in green technologies, or improvements in education and skill development. By establishing a solid framework of objectives and indicators, policymakers create a roadmap that guides transformation efforts and provides a baseline against which progress can be measured.

Data collection is pivotal in the M&E process. Accurate, timely, and relevant data enables governments and institutions to make informed decisions based on real-time economic developments. It allows for the identification of trends, the analysis of outcomes, and the comparison of performance against benchmarks. Governments can gather data through surveys, economic reports, digital tools, and consultations with industry stakeholders. Access to reliable data also facilitates transparency, helping the public and private sectors understand the impact of transformation policies and hold institutions accountable for their actions.

Continuous monitoring is essential in identifying early signs of challenges or unintended consequences. For instance, a policy to increase industrial activity might inadvertently lead to environmental degradation if sustainable practices are not adequately integrated. Similarly, rapid urbanization due to economic growth can strain infrastructure and create inequality if not managed effectively. Monitoring ensures that these issues are identified promptly so that mitigation strategies can be implemented before problems escalate. Governments can use this data to address bottlenecks, reallocate resources, or modify regulations to maintain momentum toward achieving transformation goals.

Evaluating the effectiveness of policies also helps governments assess what is working well and what needs improvement. Evaluation reports can reveal which initiatives have delivered the most substantial economic or social benefits, helping guide future investments and policy decisions. This reflection and assessment process is essential for adaptive governance, where policies are continually refined to respond to new developments or unforeseen challenges. For example, suppose a policy supporting small and medium-sized enterprises (SMEs) is found to be creating fewer jobs than anticipated. In that case, the evaluation might suggest providing additional financial assistance or improving business development services to enhance outcomes.

In addition to internal evaluations, involving external stakeholders in the M&E process is beneficial. Engaging independent organizations, think tanks, or international bodies in reviewing economic transformation initiatives can provide an objective perspective and introduce global best practices. Independent evaluations help ensure that assessment processes are more comprehensive and freer from bias, considering multiple viewpoints and expertise. Collaborating with external partners also allows for sharing knowledge, insights, and innovations that can improve the design and execution of future policies.

M&E also plays a critical role in ensuring that economic transformation aligns with the broader goals of sustainability, inclusivity, and equity. By measuring policies' social and environmental impact, governments can ensure that economic growth does not come at the expense of marginalized communities or natural ecosystems. For instance, tracking metrics related to income inequality, access to education, or renewable energy adoption can provide insight into whether transformation efforts benefit all segments of society or if corrective actions are necessary to address disparities. Social and environmental indicators are as important as economic ones in creating a holistic picture of the transformation's success.

One of the most significant benefits of M&E is its ability to foster accountability and transparency. Regularly publishing M&E reports and making data publicly accessible allows citizens, investors, and other stakeholders to track progress and hold governments accountable for their actions. Transparent reporting builds trust, promotes civic engagement, and encourages dialogue between governments and their constituents. Furthermore, transparent M&E processes can attract foreign investors who seek stability and clarity in the countries where they operate.

Monitoring and evaluation are indispensable for ensuring the success of economic transformation initiatives. By setting clear goals, collecting and analyzing data, and continuously assessing the effectiveness of policies, governments can make informed adjustments that drive progress and address challenges in real time. Data-driven decision-making ensures that transformation efforts remain adaptable, accountable, and aligned with broader objectives like sustainability, inclusivity, and long-term resilience. As the economic development landscape evolves, robust M&E frameworks will be essential for achieving meaningful and sustainable change.

9.3 Examples of Successful Economic Transformation

Successful examples of economic transformation can be found in countries like South Korea, Singapore, and Germany. South Korea shifted from an agrarian economy to a global leader in technology and manufacturing through strategic investments in education, innovation, and infrastructure. Singapore transitioned from a trade-dependent port city to a financial and technological hub, driven by solid governance and policy reforms. After World War II, Germany rebuilt its economy into a powerhouse through industrial diversification, innovation, and a focus on sustainability. These nations illustrate how strategic planning, investment in human capital, and embracing technology can drive transformative, sustainable economic growth.

9.3.1 South Korea's economic transformation: From agrarian economy to industrial powerhouse

South Korea's transformation from a primarily agrarian society into a global industrial and technological leader is one of the most remarkable economic success stories of the 20th century. In the aftermath of the Korean War (1950-1953), South Korea was one of the poorest countries in the world, with a largely rural population dependent on subsistence farming. However, through strategic investments in education, technology, and infrastructure, the country embarked on a rapid and sustained development trajectory that catapulted it into the ranks of the world's leading economies. The nation's commitment to human capital development, innovation, and infrastructure played key roles in this transformation, setting a model for other developing countries.

One of the cornerstones of South Korea's economic transformation was its heavy investment in education. Recognizing that human capital is essential for sustainable economic growth, the South Korean government focused on improving access to education at all levels, from primary schooling to higher education and vocational training. This investment paid off by creating a highly skilled and adaptable workforce supporting the country's rapid industrialization. By the 1970s, South Korea had achieved universal literacy, and its emphasis on STEM (science, technology, engineering, and mathematics) education laid the foundation for developing high-tech industries. This focus on education allowed the country to move from low-wage manufacturing to high-value sectors like electronics, automotive, and telecommunications.

Technology and innovation also played pivotal roles in South Korea's transformation. In the 1960s and 1970s, the South Korean government actively promoted the development of key industries such as steel, shipbuilding, and textiles. However, by the 1980s, it shifted its focus to more technology-driven sectors. The government provided financial and regulatory support to large conglomerates, known as chaebols, which included companies like Samsung, LG, and Hyundai. These firms invested heavily in research and development (R&D), driving innovation in consumer electronics, semiconductors, and automobiles. By fostering close cooperation between government, industry, and academia, South Korea was able to leapfrog into cutting-edge industries, establishing itself as a global leader in technology and innovation by the 1990s.

Infrastructure development was another critical factor in South Korea's economic success. After the devastation of the Korean War, the government prioritized rebuilding the country's physical infrastructure, including roads, railways, ports, and power plants. The construction of large industrial complexes, such as the Ulsan Industrial Complex, helped create the conditions for industrial growth by providing modern facilities for manufacturing and heavy industries. Additionally, South Korea invested in telecommunications and digital infrastructure, ensuring that the country could support the rapid growth of its information technology and electronics sectors. By the early 2000s, South Korea had one of the world's most advanced broadband and mobile telecommunications networks, further fueling its competitive advantage in high-tech industries.

South Korea's success also depended on its export-driven growth strategy. In the 1960s, the government shifted from import substitution policies to an outward-oriented industrialization model, focusing on producing goods for export. This strategy helped South Korean industries access global markets, generate foreign exchange, and accelerate technological advancements through competition. The government supported this approach by providing export incentives, establishing free trade zones, and negotiating trade agreements with key partners. As a result, South Korea's exports surged, becoming a major player in global markets for products like automobiles, electronics, and ships.

By the 21st century, South Korea had fully transitioned from a low-income, agrarian society to a high-income, advanced industrial economy. Its GDP per capita skyrocketed from less than $100 in the early 1960s to over $30,000 by 2020. South Korea's development was challenging, including the Asian financial crisis in 1997 and 1998. However, its strong institutions, innovative industrial policy, and resilient human capital helped the nation recover swiftly and continue its economic ascent. Today, South Korea ranks among the world's most innovative countries, with one of the highest R&D expenditures as a percentage of GDP.

South Korea's economic transformation is a testament to the power of strategic investments in education, technology, and infrastructure. South Korea achieved rapid and sustained economic growth by fostering a skilled workforce, encouraging technological innovation, and building the physical and digital infrastructure necessary for industrialization. Its successful transformation from an agrarian economy to a technological powerhouse is a valuable case study for countries seeking similar developmental trajectories. Through strategic investments in education, technology, and infrastructure, it transitioned from a primarily agrarian economy to a leading industrial and technological powerhouse (Lee, 2003).

9.3.2 Germany's Energiewende: A blueprint for sustainable economic transformation

Germany's *Energiewende* (energy transition) policy represents one of the most ambitious national efforts to transition from fossil fuel dependence to renewable energy sources, transforming the country's energy landscape while creating new industries and employment opportunities. Launched in the early 2000s, the *Energiewende* aims to reduce greenhouse gas emissions, phase out nuclear power, and increase the share of renewable energy in Germany's energy mix, particularly wind and solar power. This shift addresses environmental sustainability and drives economic transformation by fostering innovation, boosting green industries, and generating jobs in the renewable energy sector.

At the heart of the *Energiewende* is Germany's commitment to reducing its reliance on fossil fuels such as coal, oil, and natural gas, which have traditionally powered its industries. In the aftermath of the 2011 Fukushima nuclear disaster, Germany also decided to phase out its nuclear power plants, further accelerating the need for renewable alternatives. The government set ambitious targets for carbon emissions reductions and renewable energy generation, committing to achieving an 80-95% reduction in greenhouse gas emissions by 2050 compared to 1990 levels and increasing the share of renewables to 80% of electricity consumption by the same year. This has spurred significant investments in wind, solar, and biomass energy, positioning Germany as a global leader in renewable energy adoption.

The economic impact of the *Energiewende* has been substantial. The transition to renewable energy has created a booming green economy in Germany, generating hundreds of thousands of jobs in industries such as wind turbine manufacturing, solar panel production, energy storage, and grid infrastructure development, according to Pegels and Lütkenhorst (2014), Germany's renewable energy sector employed nearly 370,000 people by the early 2010s. This number continues to grow as the country advances its green energy agenda. This economic transformation has also revitalized regions previously dependent on coal mining, as new green jobs offer alternatives to workers displaced by the decline of fossil fuel industries. The *Energiewende* mitigates the environmental impact of energy consumption and ensures that economic growth is sustainable and inclusive.

In addition to job creation, Germany's focus on renewable energy has stimulated innovation in the green technology sector. The country's significant investments in research and development (R&D) have fostered breakthroughs in energy efficiency, grid management, and energy storage technologies. Germany has become a hub for green tech startups and companies working on innovative solutions to harness and store renewable energy more effectively. This focus on technological innovation has not only strengthened the competitiveness of German industries but also established Germany as a global leader in the export of renewable energy technologies.

German firms are key players in the international market for wind turbines, solar panels, and other green technologies, contributing to the country's economic growth.

Moreover, the *Energiewende* has reshaped Germany's energy infrastructure, creating a decentralized energy system that reduces dependency on large-scale fossil fuel power plants. Renewable energy sources, particularly wind and solar, are often generated locally, leading to the development of community energy projects and decentralized energy grids. This transition promotes energy security, reduces vulnerability to global energy price fluctuations, and empowers local communities to take control of their energy needs. The decentralized model also creates opportunities for small and medium-sized enterprises (SMEs) to participate in the renewable energy market, further supporting local economic development.

However, the *Energiewende* has faced challenges, including modernizing and expanding Germany's energy grid to accommodate the intermittent nature of wind and solar power. Significant investments are required to ensure that renewable energy can be stored and distributed efficiently. Additionally, the transition has led to higher energy costs for consumers in the short term, as investments in renewable infrastructure are passed on through electricity prices. Nevertheless, the long-term benefits of reducing carbon emissions, fostering innovation, and creating green jobs are expected to outweigh these initial challenges, positioning Germany as a global leader in the fight against climate change.

Germany's *Energiewende* also highlights the importance of strong government policies and public support in driving large-scale economic transformation. The German government has implemented various policy measures to support the energy transition, including feed-in tariffs that guarantee fixed payments to renewable energy producers, grants for energy efficiency projects, and financial incentives for consumers to install solar panels or energy storage systems. These policies have been instrumental in encouraging private investment in renewable energy and driving the adoption of green technologies. Public support for the *Energiewende* remains high, with widespread recognition of the need for a sustainable energy future.

Germany's *Energiewende* represents a bold and comprehensive approach to economic transformation through environmental sustainability. By shifting from fossil fuels to renewable energy, Germany has reduced its carbon footprint and created new industries, jobs, and technological innovations that contribute to long-term economic growth. The *Energiewende* demonstrates that economic transformation and environmental stewardship can go hand in hand, providing a model for other nations seeking to transition to a greener and more sustainable economy. Implemented the "Energiewende" (energy transition) policy to shift from fossil fuels to renewable energy, creating new industries and jobs in the green economy (Pegels and Lütkenhorst, 2014).

9.3.3 Rwanda's economic transformation: From agriculture to ICT and services

Rwanda's remarkable economic transformation from a predominantly agricultural economy to a growing tech and service-driven economy is one of the most compelling development stories in recent decades. Following the devastating 1994 genocide, the country faced monumental challenges, including widespread poverty, destruction of infrastructure, and deep social divisions. However, through visionary leadership, strategic reforms, and a focus on innovation, Rwanda has redefined itself as a rising hub for information and communication technology (ICT) in Africa, significantly improving governance and fostering economic growth.

The government's emphasis on ICT as a key development driver is central to Rwanda's transformation. Recognizing the potential of technology to modernize the economy, improve governance, and increase global competitiveness, the Rwandan government launched its Vision 2020 strategy, which aimed to transition the country from a low-income agrarian economy to a knowledge-based economy. Under this strategy, significant investments were made in digital infrastructure, including establishing a nationwide 4G LTE network, which now covers over 95% of the country. This connectivity has opened up new opportunities for businesses, education, and healthcare, transforming Rwanda into one of Africa's most digitally connected countries.

The focus on ICT has spurred the development of a vibrant tech ecosystem in Rwanda, particularly in the capital, Kigali. The government has collaborated with private sector players and international partners to create innovation hubs, incubators, and training programs that nurture local talent and support tech startups. For instance, the Kigali Innovation City, a major tech hub, is home to numerous tech companies, research institutions, and universities. These initiatives have attracted foreign investment and helped position Rwanda as a regional center for technology and innovation, while creating jobs and fostering entrepreneurship.

Rwanda's transformation in the tech sector is complemented by significant progress in the services industry, particularly finance, tourism, and hospitality. The government has created a conducive environment for service sector growth by improving the ease of doing business, enhancing transparency, and investing in critical infrastructure such as roads, airports, and hospitality facilities. Kigali has become a regional hub for international conferences and events, boosting tourism and associated industries. Furthermore, Rwanda's financial sector has expanded, with innovations like mobile banking and digital financial services increasing access to banking for rural and underserved populations.

In addition to its technological and service sector advancements, Rwanda's governance reforms have played a crucial role in driving economic transformation. The country has focused on improving public sector efficiency, reducing corruption, and promoting good governance, which has been instrumental in attracting foreign direct investment (FDI). Rwanda ranks among the top countries in Africa in terms of ease of doing business, according to the World Bank's *Doing Business* report. Reforms such as simplifying business registration, streamlining tax procedures, and protecting investor rights have made Rwanda an attractive destination for both local and international investors. This has boosted confidence in the country's economic potential, helping to diversify its economy and reduce reliance on agriculture.

Rwanda's governance reforms have also fostered social and economic inclusion. The government has prioritized education and healthcare to improve human capital, ensuring that the benefits of economic growth are widely shared. The expansion of universal access to education, with a particular focus on ICT and technical skills, has equipped Rwandans with the knowledge and capabilities required to participate in the emerging digital economy. The healthcare system has also been strengthened, with the implementation of community-based health insurance programs that have increased healthcare access for rural populations.

While Rwanda has made significant strides, challenges remain. Agriculture still accounts for a substantial portion of employment, and poverty reduction continues to be a priority. The government is therefore working to balance the needs of the agricultural sector with the growing demands of the tech and services industries. The goal is to ensure that rural areas, where the majority of the population resides, are not left behind in the country's march toward modernization. This includes promoting ICT-driven solutions in agriculture, such as digital platforms for farmers to access markets and information, thus integrating technology into traditional sectors.

Rwanda's journey from a predominantly agricultural economy to a growing ICT and service-oriented economy illustrates the power of strategic planning and innovation in achieving economic transformation. By focusing on digital infrastructure, fostering a business-friendly environment, and investing in human capital, Rwanda has positioned itself as a leader in Africa's technological and economic development. Although challenges remain, the country's commitment to inclusive growth, governance reforms, and sustainability offers valuable lessons for other developing nations seeking to transform their economies and improve the well-being of their citizens. Focused on ICT and services, transforming from a predominantly agricultural economy to one with a growing tech sector and improved governance (Korpela, 2018).

Economic transformation involves a multifaceted approach to shift an economy towards more diversified, innovative, and sustainable growth. It requires coordinated efforts in policy-making, infrastructure development, public-private partnerships, SME support, trade and investment facilitation, education reform, environmental sustainability, and continuous monitoring. Successful economic transformation leads to enhanced productivity, resilience, and inclusive growth, ensuring long-term prosperity for all members of society.

9.4 Green Economy

Transition to an economic model that values environmental health and social well-being. The green economy emphasizes sustainable development by valuing environmental health and social well-being alongside economic growth. It reduces environmental risks and ecological scarcities while promoting human well-being and social equity. The transition to a green economy involves integrating environmental considerations into all economic planning and decision-making, ensuring that growth is sustainable, inclusive, and resilient (Gu et al., 2019; Cheba et al., 2022, 2023).

9.4.1 The principles of a green economy

The principles of a green economy focus on sustainability, equity, and resilience. Key principles include promoting renewable energy and resource efficiency to reduce environmental impact, ensuring social inclusiveness by providing fair access to economic opportunities and fostering economic resilience through diversification and innovation. A green economy also prioritizes the preservation of natural ecosystems, recognizes the value of ecosystem services, and encourages low-carbon technologies. It seeks to balance economic growth with environmental stewardship, addressing climate change, reducing pollution, and enhancing the well-being of current and future generations while maintaining ecological integrity.

9.4.1.1 Sustainable resource management: A pathway to long-term ecological balance

Sustainable resource management is a cornerstone of environmental stewardship, ensuring that natural resources are used in ways that meet current needs without compromising the ability of future generations to meet theirs. This approach involves maximizing efficiency, minimizing waste, and adopting practices that reduce environmental degradation. Given the increasing pressures from population growth, urbanization, and industrialization, sustainable resource management is crucial for maintaining the delicate balance between human consumption and the Earth's ability to regenerate its resources.

One key aspect of sustainable resource management is the shift toward a circular economy, where the focus is on recycling and reusing materials instead of relying on the traditional linear model of "take, make, dispose." In a circular economy, products and materials are designed to be reused, repaired, and recycled, thus extending their lifespan and reducing the extraction of raw materials. This not only minimizes waste but also lessens the environmental footprint of manufacturing processes, making industries more sustainable. For instance, recycling metals, plastics, and paper significantly reduces energy consumption and greenhouse gas emissions compared to producing these materials from virgin resources.

Sustainable agriculture is another vital component of resource management. Traditional agricultural practices often lead to soil degradation, water depletion, and loss of biodiversity. In contrast, sustainable agriculture emphasizes soil health, water conservation, and biodiversity protection through practices like crop rotation, agroforestry, organic farming, and integrated pest management. These methods improve soil fertility, reduce the need for chemical inputs, and ensure long-term agricultural productivity. Additionally, promoting sustainable fisheries and responsible forest management helps protect critical ecosystems and maintain the balance of natural cycles.

Water management plays a pivotal role in sustainable resource management as freshwater availability is increasingly threatened by climate change, pollution, and over-extraction. Sustainable water management involves strategies such as rainwater harvesting, efficient irrigation systems, water recycling, and conservation measures to reduce wastage. Responsible water management also includes protecting wetlands, rivers, and lakes to maintain the ecosystem services they provide, such as water filtration, flood control, and habitat for biodiversity. Addressing water scarcity and pollution is essential for ensuring that clean, accessible water remains available to both human populations and natural ecosystems.

Moreover, sustainable resource management recognizes the importance of natural ecosystems and the services they provide. Forests, wetlands, grasslands, and oceans play critical roles in carbon sequestration, oxygen production, and maintaining biodiversity. By protecting and restoring these ecosystems, sustainable resource management helps to mitigate climate change and safeguard the planet's resilience against environmental shocks.

Responsible management of forests through reforestation and avoiding deforestation, for example, is vital for reducing carbon emissions and preserving biodiversity, which in turn supports the health and stability of global ecosystems.

In the context of energy, sustainable resource management involves the shift from fossil fuel dependence to renewable energy sources such as solar, wind, hydro, and geothermal power. This transition not only reduces carbon emissions but also limits the environmental damage caused by fossil fuel extraction and use. Promoting energy efficiency in buildings, transportation, and industry also plays a crucial role in reducing resource consumption and mitigating the effects of climate change.

Sustainable resource management is essential for ensuring that natural resources are utilized in ways that preserve their availability for future generations. By embracing principles such as recycling, sustainable agriculture, responsible water management, and ecosystem protection, societies can reduce environmental degradation, foster long-term economic growth, and contribute to global sustainability. This approach ensures that human needs are met in harmony with nature, securing the planet's health and resources for the generations to come.

9.4.1.2 Low-carbon development: A pillar of the green economy

Low-carbon development is a critical component of the green economy, aimed at reducing carbon emissions while promoting sustainable economic growth. As economies grow and industrialize, the environmental costs—particularly the release of greenhouse gases—can have long-lasting negative impacts on the planet. To mitigate these effects, low-carbon development emphasizes the shift from carbon-intensive processes, such as those driven by fossil fuels, to more sustainable alternatives like renewable energy sources, energy efficiency measures, and the deployment of low-carbon technologies.

One of the primary strategies in low-carbon development is the transition to renewable energy. Renewable energy sources such as solar, wind, hydro, and geothermal power are virtually carbon-free, offering a sustainable way to meet the world's growing energy demands without exacerbating climate change. This shift reduces dependence on fossil fuels like coal, oil, and natural gas, which are major contributors to greenhouse gas emissions. Countries worldwide are increasingly investing in renewable energy infrastructure, recognizing that these resources are not only better for the environment but also provide long-term economic benefits, such as energy security and job creation in the green energy sector.

Energy efficiency is another crucial aspect of low-carbon development. Improving energy efficiency in industries, buildings, transportation, and households reduces the overall demand for energy, thus lowering emissions. For instance, energy-efficient technologies such as LED lighting, smart thermostats, and advanced insulation systems help reduce the amount of energy consumed in daily operations. Additionally, electric vehicles (EVs) and public transportation systems designed to run on clean energy further contribute to the reduction of emissions from the transportation sector, which is one of the largest contributors to global carbon emissions.

Low-carbon technologies, including carbon capture and storage (CCS), play a significant role in the transition to a green economy. CCS technology captures carbon dioxide (CO2) emissions from industrial processes and power plants and stores them underground, preventing them from being released into the atmosphere. This technology is especially useful for industries where emissions are difficult to eliminate completely, such as cement, steel, and chemical manufacturing. By integrating CCS and other carbon-reducing technologies, economies can significantly reduce their environmental footprint while maintaining industrial productivity.

Beyond the environmental benefits, low-carbon development also creates new economic opportunities. The growth of green industries—such as renewable energy, energy-efficient manufacturing, and sustainable transportation—creates jobs, drives innovation, and fosters technological advancements. As countries transition to low-carbon pathways, they stimulate sectors that contribute to long-term economic resilience, reduce energy costs, and improve air quality, thus enhancing the overall quality of life for citizens.

Low-carbon development is a vital component of the green economy, directly addressing the challenges of climate change and environmental degradation. By reducing carbon emissions through renewable energy adoption, improving energy efficiency, and implementing low-carbon technologies, economies can achieve sustainable growth while minimizing their environmental impact. This approach not only helps mitigate climate change but also positions economies for future success in an increasingly carbon-constrained world.

9.4.1.3 Social inclusion and equity in the green economy

Social inclusion and equity are central tenets of a green economy, ensuring that the benefits of economic growth are not only environmentally sustainable but also shared fairly across all segments of society. The transition to a green economy aims to address longstanding social inequalities, reduce poverty, and create opportunities for marginalized groups, while fostering a more just and inclusive economic system. This approach recognizes that a sustainable future cannot be achieved without improving the quality of life for all, especially vulnerable communities disproportionately affected by environmental degradation and economic exclusion.

One of the key aspects of promoting social inclusion in the green economy is the creation of green jobs—employment opportunities that contribute to preserving or restoring the environment. These jobs span sectors like renewable energy, sustainable agriculture, eco-friendly manufacturing, and environmental conservation. By investing in these areas, governments and businesses can provide new employment opportunities, particularly for populations that have been historically underserved or displaced by traditional, carbon-intensive industries. Green jobs offer pathways for skill development and economic empowerment, while also contributing to the broader goals of environmental sustainability.

Social equity in the green economy also involves addressing the root causes of inequality by ensuring that economic policies prioritize the needs of disadvantaged groups. This means creating access to education, training, and financing opportunities that allow marginalized communities to participate in and benefit from the green economy. For instance, providing affordable access to renewable energy in rural or low-income areas can reduce energy poverty while simultaneously reducing greenhouse gas emissions. In this way, the green economy can serve as a vehicle for reducing disparities in income and living standards by improving access to essential services like clean energy, water, and sustainable food sources.

Supporting vulnerable communities is a critical component of social inclusion in the green economy. Vulnerable populations, such as those in low-income regions, indigenous communities, and areas highly susceptible to the impacts of climate change, often bear the brunt of environmental harm. As a result, green economy policies must be designed to protect these groups by promoting climate resilience, providing social safety nets, and ensuring their voices are heard in decision-making processes. This could involve efforts like building climate-resilient infrastructure, offering financial assistance to those affected by environmental disasters, or ensuring equitable access to natural resources.

To achieve these goals, governments and international organizations must promote inclusive policy frameworks that integrate both environmental and social objectives. This includes designing social protection programs that support workers transitioning from high-emission industries to green sectors, ensuring fair labor practices in green jobs, and encouraging inclusive governance that involves all stakeholders in the decision-making process. Additionally, policies that reduce barriers to accessing education and training in green technologies can empower individuals from all backgrounds to contribute to and benefit from the growing green economy.

Social inclusion and equity are critical to the success of a green economy. By focusing on creating green jobs, promoting social equity, and supporting vulnerable communities, the green economy offers a pathway to sustainable development that benefits all members of society. Ensuring that economic growth is both environmentally and socially sustainable is key to addressing poverty, reducing inequality, and building a more just, resilient, and inclusive future.

9.4.1.4 Ecosystem protection and restoration in the green economy

Ecosystem protection and restoration are foundational principles of the green economy, emphasizing the need to preserve and restore natural ecosystems to sustain biodiversity and the essential services they provide. Healthy ecosystems are vital for human well-being and play a critical role in regulating the climate, purifying air and water, and ensuring food security through processes such as pollination. As the world grapples with the challenges of environmental degradation, habitat loss, and climate change, integrating ecosystem protection into economic activities is paramount for achieving long-term sustainability.

One of the primary reasons for prioritizing ecosystem protection is the biodiversity they support. Biodiversity refers to the variety of life forms within a given ecosystem, including plants, animals, fungi, and microorganisms. This diversity is crucial for maintaining ecosystem resilience, enabling ecosystems to adapt to changing environmental conditions and recover from disturbances such as natural disasters or human-induced pressures. By preserving biodiversity, we ensure that ecosystems continue to function effectively, providing essential services that underpin economic activities, such as agriculture, forestry, and fisheries.

Ecosystem services—the benefits that humans derive from ecosystems—are integral to economic prosperity and human health. These services include regulating the climate, purifying water, filtering air pollutants, and providing recreational opportunities. For instance, wetlands act as natural water filters, improving water quality while also serving as critical habitats for diverse species. Forests sequester carbon, helping mitigate climate change while supporting livelihoods through timber and non-timber forest products. By recognizing the economic value of these services, businesses and policymakers can better understand the importance of protecting ecosystems as part of sustainable development strategies.

Restoration efforts are equally important in the green economy, as they focus on rehabilitating degraded ecosystems to recover their functionality and productivity. Restoration initiatives can take many forms, such as reforestation, wetland restoration, and the rehabilitation of coastal habitats like mangroves and coral reefs. These efforts not only help to reverse the damage done to ecosystems but also provide significant economic benefits. For example, restoring mangroves can protect coastal communities from storm surges, enhance fish stocks, and provide local livelihoods through sustainable tourism and fisheries. By investing in restoration, communities can achieve economic growth while fostering environmental health.

Moreover, incorporating ecosystem protection and restoration into economic planning requires a shift in policy frameworks that recognize the interconnectedness of economic activities and ecological health. Governments can implement policies that promote sustainable land use practices, incentivize conservation efforts, and regulate harmful activities such as deforestation and pollution. Integrating ecosystem considerations into sectoral policies—such as agriculture, forestry, and urban planning—ensures that economic development does not come at the expense of natural resources.

Public awareness and community engagement are also crucial for successful ecosystem protection and restoration. Involving local communities in conservation efforts ensures that their traditional knowledge and practices are recognized and utilized. Community-led initiatives can enhance the effectiveness of restoration projects, as local stakeholders often have a deep understanding of their ecosystems and can provide valuable insights into sustainable management practices.

Ecosystem protection and restoration are vital components of the green economy, essential for maintaining biodiversity and the services that ecosystems provide. By prioritizing these principles, societies can create a sustainable economic framework that recognizes the intrinsic value of nature and promotes the well-being of both people and the planet. This approach safeguards essential ecosystem services, fosters resilience against climate change, supports economic growth, and enhances overall quality of life.

9.4.2 The steps of transition to a green economy

Transitioning to a green economy involves several essential steps. First, it begins with assessing current economic structures, environmental impacts, and social dynamics to identify areas for improvement. Next, clear and measurable sustainability goals must be established to guide the transformation. This is followed by the formulation of supportive policies and regulations that promote green technologies and practices. Investment in infrastructure, such as renewable energy and efficient transportation, is crucial. Additionally, education and training programs are necessary to equip the workforce with skills for green jobs. Finally, ongoing monitoring and evaluation mechanisms ensure progress and facilitate necessary adjustments along the way.

9.4.2.1 Policy frameworks and incentives in the green economy

Governments are pivotal in establishing policy frameworks that create an enabling environment for transitioning to a green economy. By designing and implementing policies that promote sustainable practices and incentivize green technologies, governments can guide economic activities towards long-term environmental and social sustainability. One of the most effective tools in this transition is carbon pricing, which assigns a cost to carbon emissions, encouraging businesses and consumers to reduce their carbon footprint. Carbon pricing, through mechanisms like carbon taxes or cap-and-trade systems, creates a financial incentive to adopt cleaner energy sources and energy-efficient practices, helping to mitigate climate change.

In addition to carbon pricing, subsidies for renewable energy play a crucial role in supporting the growth of green technologies. Governments can provide financial support for solar, wind, hydro, and other renewable energy projects through direct subsidies, feed-in tariffs, or favorable loan terms. These incentives reduce the initial costs of transitioning to cleaner energy sources, making renewable energy more competitive with traditional fossil fuels. By investing in renewable energy, governments not only reduce reliance on polluting energy sources but also stimulate job creation in emerging industries, contributing to both environmental and economic sustainability.

Tax incentives are another powerful tool in fostering a green economy. Governments can offer tax breaks or credits to businesses and individuals who invest in energy-efficient technologies, green buildings, or sustainable agricultural practices. These incentives lower the financial barriers to adopting sustainable technologies and encourage wider participation in green initiatives. For example, businesses that invest in energy-efficient machinery or retrofitting buildings with energy-saving features can benefit from reduced tax liabilities, promoting long-term cost savings and environmental responsibility.

On the other hand, discouraging harmful activities is equally important in the policy framework for a green economy. Governments should work to phase out subsidies for fossil fuels, which currently distort energy markets by making polluting energy sources artificially cheap. Eliminating these subsidies levels the playing field for renewable energy, making green technologies more competitive and reducing overall carbon emissions. Additionally, policies aimed at preventing deforestation and protecting biodiversity are crucial in maintaining ecosystem services that are vital for both human and economic well-being. Stricter regulations on logging, land use, and habitat destruction, combined with incentives for sustainable forestry and conservation, can significantly reduce environmental degradation.

International cooperation also plays a key role in establishing effective policy frameworks. Global agreements, such as the Paris Agreement, encourage countries to adopt ambitious targets for reducing carbon emissions and transitioning to renewable energy. Governments can collaborate through these frameworks to exchange best practices, access green financing, and ensure accountability in meeting sustainability goals. Multilateral institutions and development banks can also provide financial support and technical expertise to developing nations, helping them implement green policies and accelerate their transitions.

To make these policy frameworks effective, public awareness and stakeholder engagement are essential. Governments must work closely with businesses, civil society, and local communities to ensure that green policies are inclusive and widely supported. Public awareness campaigns can educate citizens about the benefits of sustainable practices, creating a culture of environmental responsibility. Involving stakeholders in the policymaking process helps address concerns, ensures equitable outcomes, and fosters a collective commitment to a green economy.

Policy frameworks and incentives are central to driving the shift towards a green economy. By implementing carbon pricing, providing subsidies for renewable energy, offering tax incentives, and phasing out harmful subsidies, governments can guide economic activities in a sustainable direction. These efforts, coupled with global cooperation and public engagement, create the conditions necessary for achieving a green, inclusive, and resilient economy that benefits both people and the planet.

9.4.2.2 Investment in green technologies and infrastructure

Investing in green technologies and infrastructure is a cornerstone of sustainable economic growth and environmental protection. The transition to a green economy requires substantial financial commitments to clean energy, sustainable transportation, and environmentally friendly buildings. These investments not only help mitigate climate change by reducing emissions but also foster economic growth by creating jobs, driving innovation, and promoting long-term resource efficiency.

One of the most impactful areas of investment is in renewable energy infrastructure. Developing wind farms, solar power plants, and other clean energy sources is crucial for reducing dependency on fossil fuels. For example, large-scale solar farms can generate significant amounts of electricity without emitting greenhouse gases, while wind energy offers a renewable alternative to coal and natural gas. These projects not only reduce the carbon footprint of energy production but also create employment opportunities in construction, maintenance, and engineering, revitalizing local economies. Countries that prioritize investments in renewables are often able to meet their energy needs sustainably while positioning themselves as global leaders in green technology.

Beyond energy, sustainable transportation is another critical focus for green infrastructure investment. Urban centers around the world are investing in public transportation systems, electric vehicle (EV) infrastructure, and bicycle networks to reduce reliance on cars and cut emissions from fossil fuel-based transportation. By developing robust public transit systems and supporting the adoption of EVs, governments can significantly decrease traffic congestion, improve air quality, and reduce greenhouse gas emissions. For instance, cities like Copenhagen have invested heavily in cycling infrastructure, promoting sustainable mobility while enhancing residents' quality of life. Similarly, the expansion of electric vehicle charging stations helps accelerate the shift to clean transportation, contributing to both environmental goals and job creation in the green mobility sector.

Green buildings and sustainable urban development represent another essential area for investment. By constructing energy-efficient buildings and retrofitting existing ones, cities can reduce energy consumption, lower operating costs, and minimize their environmental footprint. Green buildings use materials and technologies that promote energy efficiency, water conservation, and waste reduction. For example, the incorporation of solar panels, high-performance insulation, and energy-efficient appliances in building design can substantially cut energy use and carbon emissions. Moreover, sustainable urban planning that integrates green spaces, water management systems, and sustainable architecture contributes to healthier living environments, enhancing the well-being of urban populations while ensuring long-term environmental sustainability.

Investment in green technologies also stimulates innovation across multiple sectors. From smart grids that optimize energy use to advanced water filtration systems that ensure access to clean water, green technologies pave the way for more sustainable industries. Governments can support these developments through research and development (R&D) funding, creating a pipeline of new innovations that drive economic transformation. For instance, advancements in battery storage technology enable renewable energy systems to store surplus power, making wind and solar energy more reliable and widespread. Such innovations not only promote environmental sustainability but also open up new markets, attracting investors and generating revenue.

Moreover, investing in green infrastructure aligns with the broader objectives of sustainable development. As climate change intensifies and natural resources become scarcer, economies that prioritize sustainability will be better positioned to adapt to future challenges. Green infrastructure investments can mitigate risks associated with environmental degradation, such as flooding, extreme weather events, and resource scarcity, by building more resilient communities and economies.

The economic benefits of green investments are also clear. According to numerous studies, transitioning to a low-carbon economy can generate millions of jobs globally, particularly in sectors such as renewable energy, energy efficiency, and sustainable construction. These investments not only create immediate employment opportunities but also stimulate long-term economic growth by enhancing productivity, reducing operational costs, and increasing resource efficiency.

Investing in green technologies and infrastructure is essential for reducing environmental impact and promoting sustainable economic growth. Whether through the development of renewable energy infrastructure, the expansion of sustainable transportation systems, or the construction of energy-efficient buildings, these investments offer significant environmental and economic benefits. Governments, businesses, and investors must collaborate to scale up green investments, ensuring that economic growth is both inclusive and sustainable in the face of climate challenges.

9.4.2.3 Research and innovation: A key driver for green technologies and economic growth

Supporting research and development (R&D) in green technologies and sustainable practices is essential for driving innovation and fostering long-term economic growth. As the world grapples with climate change, environmental degradation, and resource depletion, investing in R&D offers pathways to developing solutions that promote environmental sustainability while boosting productivity and competitiveness across various sectors. Governments, businesses, and academic institutions play crucial roles in funding and promoting these efforts, ensuring that new ideas can flourish and be translated into practical applications.

One of the main areas where R&D in green technologies has a significant impact is in the development of clean energy solutions. Innovations in renewable energy, such as advancements in solar panels, wind turbines, and battery storage, have drastically reduced costs and increased efficiency, making clean energy more accessible and scalable. For instance, research into photovoltaic materials has led to the creation of more affordable and efficient solar cells, while advancements in energy storage technology allow for better integration of renewable energy sources into the grid. This has not only accelerated the transition to a low-carbon economy but has also spurred the growth of the renewable energy industry, creating jobs and new business opportunities.

Green entrepreneurship is another important aspect of this transformation. By supporting startups and small businesses that focus on sustainability, governments and investors can create an ecosystem that fosters innovation. For example, providing access to venture capital, grants, or incubator programs for green entrepreneurs can help bring sustainable products and services to market more quickly. Whether through the development of eco-friendly consumer goods, sustainable agriculture technologies, or waste reduction innovations, green entrepreneurs are key drivers of the economy's transition toward more sustainable practices. Encouraging entrepreneurship also helps scale new ideas and solutions, ensuring that innovations have the necessary support to make a real impact.

Moreover, promoting scientific research in sectors such as agriculture, water management, and transportation can lead to significant improvements in efficiency and sustainability. For example, R&D in sustainable agriculture practices—such as precision farming, organic farming, and soil regeneration—can improve crop yields while minimizing environmental impact. Similarly, research into advanced water management systems can help optimize the use of water resources, addressing issues such as scarcity and pollution. These innovations not only help conserve natural resources but also boost productivity and competitiveness in the agriculture sector, which is critical for food security and economic stability.

Encouraging the adoption of best practices across various sectors is essential to ensure that innovations developed through R&D are widely implemented. Governments can play a crucial role by setting standards, offering incentives, and creating regulations that promote sustainable practices. For instance, industries can be encouraged to adopt energy-efficient manufacturing processes or reduce waste through circular economy principles.

Promoting these best practices helps industries transition to more sustainable models of production and consumption, leading to reduced environmental impact and improved long-term viability.

Public-private partnerships (PPPs) are also vital for supporting green R&D. Collaboration between governments, businesses, and academic institutions can lead to the pooling of resources, expertise, and knowledge, accelerating innovation in green technologies. For example, government funding for university research in fields such as energy efficiency or carbon capture can be complemented by private sector investment, ensuring that new technologies are developed and commercialized. By working together, these sectors can drive technological advancements that benefit both the economy and the environment.

Furthermore, green R&D can stimulate economic growth by creating new industries and markets. As sustainable technologies and practices become more widespread, they generate demand for skilled labor, new products, and services. This demand, in turn, creates jobs and drives economic diversification. For example, the growth of the electric vehicle (EV) market has led to the expansion of battery manufacturing, charging infrastructure, and software development industries, contributing to economic growth while promoting environmental sustainability.

Supporting research and innovation in green technologies and sustainable practices is essential for driving economic growth while addressing environmental challenges. Investments in R&D fuel technological advancements, promote green entrepreneurship, and encourage the widespread adoption of sustainable practices across sectors. By prioritizing research and innovation, governments and businesses can ensure that the economy's transition to sustainability is both dynamic and inclusive, leading to long-term prosperity for future generations.

9.4.2.4 Capacity building and education: Empowering a green economy through skills and knowledge

Building the skills and knowledge necessary for a green economy is a critical step in ensuring sustainable development. As the global economy shifts toward environmentally responsible practices, there is a growing need for individuals, businesses, and governments to acquire the expertise required to navigate this transformation. Capacity building and education play a pivotal role in equipping the workforce with the necessary tools to innovate, adopt, and implement green technologies and sustainable practices, as well as raising public awareness of environmental challenges and solutions.

Integrating sustainability into educational curricula is one of the foundational steps in building a green economy. By incorporating environmental issues and sustainability concepts into school programs at all levels, students can develop an early understanding of the ecological challenges facing the world and the role they can play in addressing them. This approach not only fosters a sense of environmental stewardship but also prepares the next generation for green careers. Schools and universities can offer specialized courses in areas such as renewable energy, sustainable agriculture, circular economy principles, and climate science, ensuring that students gain the technical knowledge needed for a wide range of green industries.

Vocational training for green jobs is equally important, as the demand for skills in areas such as renewable energy, sustainable construction, and eco-friendly manufacturing continues to grow. Training programs can help workers transition from traditional industries to emerging sectors like solar and wind energy, electric vehicle (EV) production, or energy-efficient building design. Governments, educational institutions, and businesses can collaborate to provide vocational training and certification programs that equip workers with practical skills for these green jobs. By focusing on hands-on experience and technical proficiency, such programs can ensure that the workforce is ready to support the green economy's expansion.

Additionally, continuing education and lifelong learning initiatives are essential for ensuring that the existing workforce remains adaptable to the evolving demands of a green economy. As industries become more reliant on cutting-edge green technologies, workers will need ongoing training to stay current with new developments. Lifelong learning programs, such as online courses, workshops, and certifications, can help workers continually upgrade their skills in fields like energy efficiency, waste management, and sustainable supply chains.

Public awareness campaigns play a crucial role in educating society about environmental issues and promoting sustainable lifestyles. Raising awareness of climate change, resource depletion, and pollution can inspire individuals and communities to adopt more sustainable behaviors, such as reducing waste, conserving energy, and supporting eco-friendly products. Governments and non-governmental organizations (NGOs) can lead campaigns to educate the public on the importance of environmental sustainability, offering practical advice on how people can reduce their ecological footprint. For instance, initiatives that promote recycling, energy conservation, and responsible consumption can contribute to the collective effort of building a greener society.

Moreover, education for decision-makers and policymakers is crucial for ensuring that environmental concerns are prioritized in governance and economic planning. By offering training programs and workshops on sustainability, governments can empower policymakers to make informed decisions that align with the principles of a green economy. This may include educating public officials on the benefits of renewable energy, sustainable urban planning, and green infrastructure investments. Knowledgeable leaders are better equipped to create policies and regulations that promote environmental stewardship and guide the country toward long-term sustainability.

Capacity building and education are essential pillars for the successful transition to a green economy. By integrating sustainability into formal education, providing vocational training for green jobs, raising public awareness, and educating policymakers, societies can equip themselves with the skills and knowledge necessary to drive the green transformation. These efforts ensure that individuals at all levels are prepared to contribute to and benefit from a more sustainable future, fostering innovation, resilience, and equity in the global economy.

9.4.3 Examples of green economy initiatives

Examples of green economy initiatives include Germany's Energiewende policy, which focuses on transitioning from fossil fuels to renewable energy sources, creating green jobs, and reducing carbon emissions. In Costa Rica, a national effort to become carbon-neutral by investing in renewable energy and reforestation has positioned the country as a global leader in sustainability. Kenya has promoted green jobs through its robust investment in geothermal energy, contributing to cleaner energy and economic growth. The European Union's Green Deal aims to make the region climate-neutral by 2050, fostering innovation in green technologies and sustainable practices across member states.

9.4.3.1 Renewable energy projects: Denmark and Germany's success in wind and solar power

Countries like Denmark and Germany have been pioneers in the transition to renewable energy, particularly in the development of wind and solar power. Their commitment to reducing carbon emissions while creating sustainable jobs has positioned them as global leaders in the renewable energy sector. Through strategic investments and supportive government policies, both nations have demonstrated that it is possible to achieve economic growth while mitigating the effects of climate change.

Denmark, often cited as a model for wind energy development, has invested heavily in wind power since the 1970s. By leveraging its coastal geography, Denmark has harnessed wind resources to produce a significant portion of its energy needs. Today, Denmark generates nearly half of its electricity from wind, reducing its reliance on fossil fuels and lowering greenhouse gas emissions. This shift has not only helped Denmark meet its climate goals but has also created thousands of jobs in wind turbine manufacturing, maintenance, and installation. Denmark's wind energy industry has become a major exporter of wind technology and expertise, further contributing to its economic growth.

Germany, meanwhile, has pursued a broader approach through its Energiewende (energy transition) policy, which focuses on shifting the country's energy supply from fossil fuels to renewable sources like wind and solar power. The German government has incentivized renewable energy development through feed-in tariffs, subsidies, and tax incentives, encouraging investment in green technologies. As a result, Germany has become one of the world's leading producers of solar energy, despite its relatively low levels of sunlight compared to other regions. In addition, Germany's wind energy sector is among the largest in the world, contributing significantly to the country's energy mix.

These investments in renewable energy have also generated substantial employment in Germany. As of 2021, Germany's renewable energy sector employed over 300,000 people in areas such as engineering, manufacturing, research, and project management. The growth of green jobs has been particularly significant in regions that previously relied on coal and other fossil fuels, helping to mitigate the economic impact of phasing out carbon-intensive industries.

Both Denmark and Germany demonstrate the economic and environmental benefits of transitioning to renewable energy. By investing in wind and solar power, they have reduced carbon emissions, lessened their dependency on fossil fuels, and created resilient economies centered on green jobs and technological innovation. Their successful strategies serve as examples for other countries seeking to transition to more sustainable energy systems, proving that the path to a green economy is not only feasible but also economically rewarding. Countries like Denmark and Germany have invested heavily in wind and solar power, reducing carbon emissions and creating green jobs in the renewable energy sector (Lipp, 2007).

9.4.3.2 Sustainable agriculture: Enhancing productivity while minimizing environmental impact

Sustainable agriculture aims to balance increased productivity with the need to minimize environmental degradation, reduce greenhouse gas emissions, and promote resilience to climate change. Approaches such as organic farming, agroforestry, and precision agriculture are key methods for achieving this balance. These techniques not only enhance the ecological sustainability of farming but also improve soil health, conserve water, and increase biodiversity. Rwanda's Land Husbandry, Water Harvesting, and Hillside Irrigation projects serve as a prime example of how sustainable agricultural practices can transform an agricultural sector and adapt to the challenges posed by climate change.

Organic farming is one of the foundational approaches to sustainable agriculture. By avoiding synthetic pesticides and fertilizers, organic farming maintains soil fertility and promotes biodiversity, while reducing pollution and carbon emissions. Although organic yields may sometimes be lower than conventional farming, the ecological benefits and long-term sustainability make it a valuable practice. Organic farming also supports healthier ecosystems, reducing the dependency on external inputs, and fostering more resilient agricultural systems that can better withstand climate-related stresses.

Agroforestry, the integration of trees and shrubs into agricultural landscapes, has emerged as another essential practice in sustainable agriculture. By combining trees with crops and livestock, agroforestry enhances biodiversity, improves water retention, and prevents soil erosion. It also contributes to carbon sequestration, playing a crucial role in mitigating climate change. In Rwanda, agroforestry has been instrumental in rehabilitating degraded land and providing additional sources of income for farmers through the sale of timber, fruit, and other tree products. Agroforestry also supports improved microclimates, which benefit both the crops and the wider environment.

Precision agriculture represents a modern, technology-driven approach to sustainability. This method involves the use of advanced tools such as sensors, GPS, and data analytics to optimize farming practices. Precision agriculture allows farmers to apply inputs like water, fertilizers, and pesticides with great accuracy, reducing waste and minimizing environmental impact. For instance, by using GPS-guided irrigation systems, farmers can ensure that crops receive the exact amount of water needed, preventing overuse and conserving this precious resource. Such precision is vital for both boosting yields and protecting ecosystems.

Rwanda's Land Husbandry, Water Harvesting, and Hillside Irrigation projects, launched to address the challenges of soil erosion, water scarcity, and climate vulnerability, illustrate how sustainable agriculture can transform communities. Rwanda's steep terrain and reliance on rain-fed agriculture made its farmers particularly vulnerable to soil degradation and erratic rainfall patterns caused by climate change. The government, with support from international development agencies, implemented these projects to promote sustainable farming practices, such as terracing, water harvesting, and irrigation, on hillsides and other marginal lands.

The results have been impressive. The Land Husbandry project has not only reduced soil erosion but also improved agricultural productivity. Through hillside irrigation, farmers are now able to cultivate crops throughout the year, even during dry seasons, thereby improving food security and income stability. The use of water harvesting techniques has also enhanced the efficiency of water use, reducing vulnerability to droughts and supporting better crop yields.

Sustainable agriculture offers a holistic approach to balancing productivity with environmental stewardship. Whether through organic farming, agroforestry, or precision agriculture, these practices are essential for achieving long-term agricultural sustainability. Rwanda's success in applying sustainable land management techniques demonstrates the potential of these approaches to transform agriculture in ways that support both economic development and ecological resilience. By focusing on sustainability, countries can build more robust and climate-resilient agricultural sectors, ultimately contributing to global efforts to combat climate change while ensuring food security for future generations. Organic farming, agroforestry, and precision agriculture help increase productivity while reducing environmental impact. For example, Rwanda's Land Husbandry, Water Harvesting, and Hillside Irrigation projects have improved agricultural productivity and resilience to climate change (Mundere and Marešová, 2016).

9.4.3.3 Green urban development: Sustainable cities for a healthier future

Green urban development is essential for creating cities that are not only livable but also environmentally sustainable. Cities like Curitiba in Brazil and Copenhagen in Denmark serve as prime examples of how sustainable urban planning can enhance the quality of life for residents while minimizing environmental impact. These cities have implemented a range of initiatives, including public transportation networks, green spaces, and energy-efficient buildings, that contribute to both ecological preservation and social well-being.

Curitiba is a global model for green urban development. Since the 1960s, the city has prioritized sustainability in its urban planning, particularly through its renowned public transportation system. Curitiba's Bus Rapid Transit (BRT) system, a pioneering initiative, offers an affordable, efficient, and low-carbon public transportation option that reduces traffic congestion and air pollution. The BRT system is designed to encourage public transit use over private vehicles, which cuts down on emissions and creates a more accessible, connected city. In addition, Curitiba has integrated extensive green spaces into its cityscape. Parks, gardens, and open spaces serve as natural flood control systems while providing recreational areas for residents. These green zones improve air quality, preserve biodiversity, and enhance the city's resilience to climate change.

Copenhagen is another leading city in sustainable urban development, known for its ambitious efforts to become the world's first carbon-neutral capital by 2025. The city's comprehensive approach includes promoting bicycle-friendly infrastructure, energy-efficient buildings, and renewable energy use. Copenhagen's extensive network of cycling lanes and its focus on pedestrian-friendly urban design significantly reduce car usage, contributing to lower emissions and better air quality. Additionally, the city has invested in energy-efficient buildings through strict green building codes and retrofitting older structures to improve insulation and energy usage. These efforts are supported by the city's use of renewable energy, such as wind and solar power, which fuels a significant portion of Copenhagen's energy needs.

Both cities exemplify how green urban development can promote sustainability and improve urban living conditions. Curitiba and Copenhagen's focus on energy-efficient buildings also highlights the importance of green architecture in reducing the environmental impact of urban areas. Energy-efficient buildings incorporate features like high-performance insulation, energy-efficient heating and cooling systems, and the use of renewable energy sources such as solar panels. These buildings not only lower energy consumption but also contribute to reducing carbon footprints.

Green spaces are another key element in these cities' urban planning. In addition to offering recreational and aesthetic benefits, they play a crucial role in urban ecosystems. Trees and plants in these areas act as carbon sinks, improving air quality, cooling the urban environment, and mitigating the heat island effect. Moreover, green spaces promote mental and physical well-being among residents, offering a place for relaxation, exercise, and social interaction.

Copenhagen has also embraced green energy as part of its urban planning strategy. The city's commitment to renewable energy is evident in its widespread use of wind power, which supplies a substantial portion of the city's electricity needs. Additionally, Copenhagen's focus on district heating, which distributes heat from a centralized location to residential and commercial buildings, allows for more efficient energy use and reduced emissions. This combination of renewable energy and energy-efficient infrastructure positions Copenhagen as a global leader in sustainable urban development.

Green urban development offers a path to creating cities that are not only economically and socially vibrant but also environmentally sustainable. Curitiba and Copenhagen illustrate how sustainable urban planning—through public transportation, green spaces, energy-efficient buildings, and renewable energy—can reduce environmental impact while enhancing the quality of life for residents. As cities around the world continue to grow, adopting these green development practices will be essential for addressing climate change, reducing resource consumption, and building more resilient urban communities. Cities like Curitiba in Brazil and Copenhagen in Denmark have implemented sustainable urban planning initiatives, including extensive public transportation networks, green spaces, and energy-efficient buildings, to enhance the quality of life for residents while reducing environmental impact (Gürçam, 2024).

9.4.3.4 Circular economy models: Redefining resource use for a sustainable future

The circular economy represents a transformative shift in how economies manage resources, focusing on maximizing the lifecycle of materials through reuse, recycling, and regeneration. Unlike the traditional linear economy, where goods are produced, consumed, and disposed of, the circular model ensures that resources are kept in use for as long as possible. This shift minimizes waste, conserves resources, and creates opportunities for innovation and economic growth.

A key example of this transformation is the European Union's Circular Economy Action Plan. Launched as part of the European Green Deal, this initiative aims to transition Europe from a linear to a circular economy. The Action Plan addresses multiple sectors, including electronics, textiles, plastics, and construction, and introduces measures to promote the reuse, repair, and recycling of products. By designing products with a longer lifespan and encouraging the remanufacturing of goods, the plan seeks to reduce waste and lessen Europe's dependence on raw materials.

The principles of the circular economy include designing out waste and pollution, keeping products and materials in use, and regenerating natural systems. This model involves extending the life cycle of products through various processes such as repair, refurbishment, and recycling. For example, instead of discarding electronic devices once they become outdated, a circular economy encourages the refurbishment of these items, allowing them to be reused or upgraded. This not only reduces waste but also fosters innovation in product design, as companies are incentivized to create goods that can be easily repaired or recycled.

One of the most critical aspects of the circular economy is its focus on closing resource loops. In a circular system, the materials used in products are fed back into the production cycle at the end of their life. This reduces the need for extracting new raw materials, which is especially important for finite resources like metals and minerals. Recycling processes ensure that materials like plastics, metals, and textiles can be recovered and repurposed, thus reducing the environmental impact of resource extraction and processing.

The EU's Circular Economy Action Plan also emphasizes eco-design, where products are created with sustainability in mind. This includes designing products that require fewer resources to manufacture, consume less energy during use, and are easier to disassemble for recycling at the end of their lifecycle. By encouraging such design practices, the circular economy fosters the development of products that have minimal environmental impact.

Circular economy models not only offer environmental benefits but also create new economic opportunities. By shifting towards a system where materials and products are continuously reused, businesses can lower production costs, reduce resource dependency, and explore new markets for refurbished goods and recycling services. The circular economy also promotes job creation in sectors like recycling, remanufacturing, and sustainable product design. For instance, industries involved in recycling and waste management benefit from increased demand for recycled materials, while technology companies can generate revenue by offering refurbishment and upgrade services for electronics.

Consumer behavior also plays a crucial role in the success of a circular economy. Educating consumers about sustainable consumption, encouraging the sharing and leasing of products, and promoting responsible disposal practices are vital for maintaining circular systems. The shift towards a circular economy requires a cultural change in how society views ownership and consumption, moving from a throwaway culture to one that values sustainability and resource efficiency.

Overall, the transition to a circular economy is essential for achieving long-term sustainability. By reducing waste, conserving natural resources, and fostering economic resilience, circular models present a viable solution to the environmental and economic challenges posed by resource depletion and pollution. The EU's commitment to the circular economy sets an example for other regions, demonstrating how innovative policies and practices can create a more sustainable and prosperous future. The European Union's Circular Economy Action Plan aims to transition from a linear economy (take, make, dispose) to a circular one, where resources are reused, recycled, and kept in the economy for as long as possible. This approach reduces waste, conserves resources, and creates new economic opportunities (Lewandowski, 2016).

9.4.4 The benefits of transitioning to a green economy

Transitioning to a green economy brings a multitude of benefits that extend across environmental, economic, and social dimensions. A green economy prioritizes sustainability, reducing the environmental impact of economic activities while promoting economic growth and improving societal well-being. The shift to greener practices not only mitigates the threats posed by climate change but also offers pathways for innovation, job creation, and long-term economic resilience. Here are some key benefits of transitioning to a green economy:

9.4.4.1 Environmental protection and climate mitigation

One of the most significant advantages of a green economy is the reduction of environmental degradation and the mitigation of climate change. By adopting renewable energy sources like wind, solar, and hydropower, a green economy reduces greenhouse gas emissions and decreases dependency on fossil fuels. Sustainable practices, such as resource efficiency, waste reduction, and the protection of ecosystems, help preserve biodiversity and reduce pollution. As countries transition to a green economy, they contribute to meeting global climate goals, such as those outlined in the Paris Agreement, thereby playing a role in stabilizing the global climate and protecting natural resources for future generations.

9.4.4.2 Economic growth and job creation

Contrary to the misconception that environmental policies hinder economic growth, transitioning to a green economy can actually stimulate new industries and create jobs. The growth of sectors like renewable energy, sustainable agriculture, waste management, and green construction offers substantial opportunities for employment. For example, the renewable energy industry is already one of the fastest-growing sectors worldwide, employing millions in areas such as solar panel installation, wind turbine manufacturing, and energy storage. Furthermore, the circular economy, which focuses on recycling, refurbishing, and reusing products, opens avenues for innovation and entrepreneurship, creating jobs in design, remanufacturing, and resource management. This shift fosters economic diversification, making economies more resilient to external shocks.

9.4.4.3 Resource efficiency and cost savings

A green economy emphasizes resource efficiency, which reduces costs for businesses and governments over time. Through sustainable practices like recycling, energy conservation, and water management, industries can lower production costs while minimizing waste. For example, energy-efficient buildings and manufacturing processes reduce energy consumption, cutting operational expenses and making businesses more competitive. Governments that invest in green infrastructure, such as sustainable transportation and renewable energy grids, also reduce long-term costs associated with environmental damage, health care from pollution, and climate-related disasters. In the long run, resource efficiency enhances economic stability and sustainability.

9.4.4.4 Health and well-being

The green economy has the potential to significantly improve public health and overall well-being. Reducing air, water, and soil pollution through sustainable practices can lead to lower incidences of diseases like respiratory problems, heart disease, and cancer, which are often linked to environmental pollutants. For instance, transitioning to clean energy reduces air pollution, benefiting public health by lowering the risk of respiratory illnesses caused by fossil fuel emissions. Moreover, sustainable urban planning that prioritizes green spaces and energy-efficient housing improves the quality of life for citizens, fostering a healthier and more sustainable living environment.

9.4.4.5 Social equity and inclusivity

A green economy is inherently focused on social inclusion and equity. The transition to sustainable practices offers an opportunity to address systemic inequalities, providing marginalized communities with better access to clean energy, green jobs, and healthier environments. Governments can implement policies that ensure the fair distribution of green economy benefits, such as job training programs for vulnerable populations or subsidies for energy-efficient technologies in low-income households. Additionally, green jobs and sustainable infrastructure projects can uplift rural areas and communities that have traditionally been left behind by industrial growth, offering them new opportunities for participation in the economy.

9.4.4.6 Energy security and independence

By investing in renewable energy sources, countries can reduce their dependency on imported fossil fuels, leading to greater energy security and independence. Nations that transition to wind, solar, and other renewable energy technologies are less vulnerable to global energy price fluctuations, political instability in oil-producing regions, or supply chain disruptions. Green energy systems also promote decentralized energy production, allowing communities to produce their own energy locally, which enhances resilience and autonomy. In the long term, green energy infrastructure strengthens national security and stabilizes energy supplies.

9.4.4.7 Resilience to future economic and environmental shocks

The transition to a green economy prepares countries to be more resilient in the face of future economic and environmental shocks. Countries can buffer themselves against the volatility of fossil fuel markets and resource depletion by reducing dependence on finite natural resources. Furthermore, investments in sustainable infrastructure and green technologies make economies more adaptable to climate-related disruptions, such as extreme weather events, rising sea levels, and resource scarcity. Building a green economy that prioritizes sustainability ensures that societies are better equipped to face the challenges of the 21st century and beyond.

It is a critical step toward a more sustainable, inclusive, and prosperous future. By addressing environmental challenges, promoting innovation, and creating economic opportunities, the green economy offers a pathway for long-term growth and resilience. Governments, businesses, and individuals must collaborate to accelerate this transition, ensuring that the benefits of a green economy are shared equitably across society and that future generations can enjoy a healthy planet and a thriving economy.

A green economy is an economic model that prioritizes environmental health and social well-being alongside financial growth. The transition to a green economy involves sustainable resource management, low-carbon development, social inclusion, ecosystem protection, and the implementation of supportive policy frameworks and incentives. By investing in green technologies, fostering innovation, building capacity, and encouraging public-private partnerships, societies can achieve sustainable, inclusive, and resilient growth.

9.5 Sustainable Finance

Invest in projects and businesses prioritizing sustainability and ethical practices. Sustainable finance refers to integrating environmental, social, and governance (ESG) criteria into financial services, including investment decisions, lending, and insurance. It aims to support economic growth while reducing environmental pressures, tackling social issues, and improving governance. Sustainable finance involves directing capital toward projects and businesses, prioritizing sustainability and ethical practices, and ensuring that financial flows contribute to long-term societal benefits (Sepetis et al., 2022; Mikelsone et al., 2021; Ferri and Lipari, 2019).

The importance of sustainable finance lies in its potential to drive the transition to a more sustainable economy. Channeling investments into sustainable projects encourages businesses to adopt greener practices, innovate in sustainable technologies, and promote social equity. Sustainable finance helps mitigate risks associated with environmental degradation and social unrest, creating more resilient economies and societies.

9.5.1 Key components of sustainable finance

Sustainable finance integrates environmental, social, and governance (ESG) factors into financial decision-making. Key components include green bonds, which fund environmentally friendly projects, and impact investing, aiming for both financial returns and positive social or environmental outcomes. Climate risk assessment evaluates financial risks linked to climate change, ensuring resilience. Sustainable banking promotes green loans and responsible investments, while corporate responsibility aligns business strategies with sustainability goals. Together, these elements foster a financial system that supports long-term environmental stewardship, social equity, and economic growth.

9.5.1.1 Green bonds

Green bonds are financial instruments designed to support projects with positive environmental outcomes. These fixed-income securities offer investors a way to contribute to sustainable development while earning a return. Issuers, typically governments, municipalities, or corporations, raise capital specifically for climate and environmental projects, making green bonds a key tool in addressing global environmental challenges such as climate change, resource depletion, and pollution.

One of the primary areas where green bonds are employed is in financing renewable energy installations, such as wind farms, solar power plants, and hydroelectric projects. These investments help reduce dependency on fossil fuels, mitigate carbon emissions, and promote cleaner energy production. In doing so, green bonds not only advance sustainability but also support energy transitions and innovation in the renewable sector.

Additionally, energy efficiency improvements in buildings and industrial facilities are commonly funded by green bonds. Retrofitting old buildings, upgrading heating and cooling systems, and improving insulation all contribute to reduced energy consumption and lower greenhouse gas emissions. By directing funds to these initiatives, green bonds facilitate more sustainable urban development and business practices, reducing overall environmental footprints.

Green bonds also fund clean transportation initiatives, including electric vehicle infrastructure, public transit improvements, and the development of low-emission transportation networks. This shift toward sustainable mobility can reduce air pollution, decrease traffic congestion, and promote healthier urban environments.

Moreover, green bonds are used in sustainable water management and conservation efforts, supporting projects like water treatment facilities, flood management systems, and biodiversity preservation. These initiatives enhance the sustainable use of natural resources and protect ecosystems vital to human survival.

By channeling investments into these sectors, green bonds create opportunities for investors to contribute to the global shift toward sustainability while benefiting from secure, fixed-income returns.

These are fixed-income securities specifically earmarked to raise money for climate and environmental projects. Green bonds can finance various projects, including renewable energy installations, energy efficiency improvements, clean transportation, sustainable water management, and conservation efforts.

9.5.1.2 Social bonds

Social bonds are financial instruments aimed at funding projects that generate positive social outcomes, addressing critical societal issues while providing investors with returns. Unlike traditional bonds, social bonds focus on fostering social equity, inclusion, and well-being. The capital raised through social bonds is specifically earmarked for initiatives that benefit marginalized or underserved communities, making them a vital tool in advancing social development alongside economic progress.

One key area where social bonds are utilized is in affordable housing. Many governments and private institutions issue social bonds to fund the construction or renovation of housing for low-income individuals and families. This helps address housing shortages, reduces homelessness, and provides stable living conditions, which are essential for social and economic well-being. Access to affordable housing also contributes to community stability, supporting employment and educational opportunities for residents.

Education is another sector where social bonds make a significant impact. Funds raised through social bonds can be allocated to improving school infrastructure, expanding access to quality education in underserved areas, and developing vocational training programs that equip individuals with skills needed in the workforce. By enhancing educational opportunities, social bonds promote long-term social mobility and economic growth.

In the realm of healthcare, social bonds are instrumental in financing projects that provide better access to medical services, particularly for vulnerable populations. This includes building hospitals, expanding telemedicine services, and supporting preventive care initiatives. Improved healthcare access leads to healthier communities, reducing the burden of diseases and enhancing productivity.

Furthermore, social bonds support employment generation and entrepreneurship among marginalized groups, such as women, refugees, and individuals with disabilities. By funding job training programs, microfinance initiatives, and inclusive business models, social bonds help to break cycles of poverty and empower disadvantaged individuals to participate fully in the economy.

Social bonds represent a growing financial mechanism that aligns the goals of social impact with financial returns, addressing critical societal challenges like inequality, poverty, and limited access to essential services. Through these bonds, governments, corporations, and development institutions can channel capital toward creating more inclusive, resilient societies.

9.5.1.3 Sustainable investment funds

Sustainable Investment Funds are financial instruments designed to align investment strategies with environmental, social, and governance (ESG) criteria. These funds prioritize sustainability in their portfolio selection, focusing on companies and sectors that uphold strong ESG standards while excluding industries or activities that harm society and the environment, such as fossil fuels, tobacco, or weapons. By investing in companies that lead in sustainable practices, these funds allow investors to support responsible corporate behavior while seeking competitive financial returns.

One of the key features of sustainable investment funds is their emphasis on environmental sustainability. These funds often prioritize companies that are actively reducing their carbon footprint, adopting renewable energy, or engaging in other green technologies. For instance, investments may be directed toward companies in the clean energy sector, such as solar and wind energy providers, or in industries focusing on sustainable agriculture, waste reduction, and water conservation. This approach not only supports the global transition toward a low-carbon economy but also mitigates risks associated with climate change and resource depletion, which can impact long-term profitability.

Social criteria play a crucial role in sustainable investment funds as well. Companies that foster diversity, promote fair labor practices, ensure employee well-being, and engage with their local communities are favored in these portfolios. By focusing on socially responsible businesses, sustainable funds contribute to addressing inequalities and improving the overall quality of life, while also tapping into markets that are increasingly valuing corporate social responsibility.

The governance aspect ensures that companies within these funds maintain transparency, accountability, and sound corporate practices. This includes strong leadership, ethical business practices, and effective stakeholder engagement. Companies with solid governance are generally seen as less prone to risks like fraud, corruption, or mismanagement, making them more stable and attractive for long-term investment.

For investors, sustainable investment funds provide an opportunity to align their portfolios with personal values and global sustainability goals. They offer a way to support businesses that contribute to a healthier planet and society, while also potentially benefiting from competitive financial returns. As consumers and regulators demand more ethical and environmentally friendly business practices, companies with strong ESG credentials are likely to see continued growth, making sustainable investment funds a promising choice for future-focused investors.

9.5.1.4 Impact investing

Impact Investing refers to a strategic investment approach where financial capital is directed towards companies, organizations, or funds with the explicit intention of generating measurable social and environmental impact, alongside financial returns. Unlike traditional investments, which prioritize profit maximization, impact investors seek to achieve positive outcomes in areas such as poverty reduction, renewable energy development, or enhanced access to clean water, while also securing financial gains.

Impact investing spans a wide range of sectors, including renewable energy, sustainable agriculture, affordable housing, education, and healthcare. For instance, investments in solar energy companies not only contribute to reducing carbon emissions but also provide clean, reliable power to underserved communities. Similarly, investing in companies focused on affordable housing or access to education creates tangible improvements in living standards and educational opportunities, especially for marginalized populations.

A key feature of impact investing is the commitment to measurable impact. Investors track and report on the social and environmental outcomes of their investments, using standardized metrics such as the Global Impact Investing Network's (GIIN) Impact Reporting and Investment Standards (IRIS). This accountability ensures that investments are producing the intended positive outcomes, such as a reduction in poverty or increased access to essential resources like clean water and healthcare.

Impact investing appeals to a growing number of investors who want to make a difference while generating returns. Targeting social and environmental objectives offers a way to address global challenges like climate change, inequality, and resource scarcity. As more investors seek purpose-driven opportunities, impact investing is expected to play an increasingly significant role in shaping a sustainable and equitable future.

9.5.1.5 ESG integration

ESG Integration is the process of incorporating environmental, social, and governance (ESG) factors into traditional financial analysis and investment decision-making. This approach acknowledges that ESG issues, such as climate change, labor practices, and corporate governance, can have material financial impacts on a company's performance and risk profile. By integrating these factors, investors gain a more comprehensive understanding of both risks and opportunities that might not be evident through conventional financial metrics alone.

In environmental terms, companies are assessed based on their sustainability practices, resource efficiency, and efforts to reduce carbon footprints. For instance, firms that adopt clean energy solutions or implement resource-saving measures may be viewed as less risky and better positioned for long-term success in a world shifting towards sustainability.

From a social perspective, ESG integration examines factors like employee welfare, diversity and inclusion, community relations, and human rights. Companies that uphold strong social standards are often more resilient to reputational risks and better able to attract talent, enhancing their long-term competitiveness.

Governance is the third pillar, focusing on how companies are managed, including their board structure, executive compensation, transparency, and ethical practices. Strong governance frameworks are critical for ensuring accountability and mitigating risks related to corruption, fraud, or mismanagement.

By incorporating ESG criteria, investors can align their portfolios with responsible business practices while potentially avoiding long-term financial risks. ESG integration not only enhances decision-making but also reflects a growing trend toward responsible investing, where financial returns are pursued alongside positive societal outcomes.

9.5.1.6 Sustainable banking

Sustainable Banking is a financial approach that integrates sustainability principles into banking operations, lending practices, and investment portfolios. This model emphasizes environmental and social responsibility alongside financial profitability. By adopting sustainable banking, institutions contribute to the global push toward sustainable development, promoting long-term ecological and social well-being.

One key aspect of sustainable banking is offering green loans. These are loans specifically designed to fund environmentally-friendly projects, such as renewable energy installations, energy-efficient infrastructure, and sustainable agriculture. By providing favorable financing for such projects, banks incentivize businesses and individuals to adopt greener practices.

Another important element is financial inclusion. Sustainable banks focus on providing access to financial services for underserved or marginalized communities, helping to reduce poverty and promote equitable economic growth. This involves offering microfinance, low-interest loans, and financial literacy programs to empower these groups economically.

Moreover, banks are increasingly focused on reducing their own carbon footprint. This involves transitioning to energy-efficient technologies in their offices, reducing paper usage, and promoting digital banking. By adopting eco-friendly practices, banks can lead by example and demonstrate their commitment to sustainability.

Sustainable banking also includes aligning investment portfolios with environmental, social, and governance (ESG) criteria. Banks and financial institutions prioritize investments in companies and projects that adhere to high sustainability standards while avoiding industries with harmful practices like fossil fuels or deforestation.

9.5.2 The steps of sustainable finance

Sustainable finance works by integrating environmental, social, and governance (ESG) factors into financial decision-making processes. The steps include identifying and evaluating investments that prioritize sustainability, such as green bonds, social bonds, or impact investing. Financial institutions set ESG criteria to assess the sustainability performance of companies or projects, while aligning portfolios with these values. Sustainable finance also promotes responsible lending and investment practices, such as funding renewable energy, sustainable infrastructure, and inclusive economic development. Continuous monitoring ensures that investments generate measurable environmental and social impact, contributing to long-term financial resilience and global sustainability goals.

9.5.2.1 Setting standards and frameworks

Setting Standards and Frameworks is a foundational step in the development of sustainable finance, ensuring that financial institutions operate transparently and responsibly. To achieve this, global initiatives like the Principles for Responsible Investment (PRI), the Equator Principles, and the Task Force on Climate-related Financial Disclosures (TCFD) have emerged as critical tools. These frameworks guide financial institutions in incorporating environmental, social, and governance (ESG) criteria into their decision-making, making sustainability a measurable and actionable priority.

The Principles for Responsible Investment (PRI), supported by the United Nations, encourage investors to commit to integrating ESG issues into their investment processes. By adhering to PRI, investors gain a framework for considering sustainability risks and opportunities across their portfolios. This standard promotes long-term thinking and encourages transparency, pushing institutions to report on their ESG activities and impacts.

The Equator Principles, on the other hand, focus on financial institutions providing project financing, particularly in high-risk sectors such as infrastructure and energy. These principles establish a risk management framework that ensures banks and other lenders evaluate the environmental and social risks of the projects they finance. By adhering to these principles, financial institutions can minimize harm to the environment and local communities, while aligning with global sustainability goals.

The Task Force on Climate-related Financial Disclosures (TCFD) emphasizes the importance of transparent reporting on climate-related risks and opportunities. Established by the Financial Stability Board, TCFD encourages companies and financial institutions to disclose how climate change impacts their financial performance. This framework helps investors understand how companies are addressing climate risks and integrating mitigation strategies into their operations.

By adopting these standards and frameworks, financial institutions can align their practices with global sustainability goals. These tools also facilitate accountability and comparability by providing clear guidelines for assessing ESG performance. As a result, stakeholders—including investors, regulators, and consumers—gain greater confidence in the integrity of sustainable finance initiatives. Ultimately, these frameworks serve as a cornerstone for scaling sustainable finance, allowing it to become a driver of global economic transformation that prioritizes both financial returns and societal well-being.

9.5.2.2 Risk management

Risk Management is a fundamental aspect of sustainable finance that involves integrating Environmental, Social, and Governance (ESG) factors into traditional risk assessment processes. As global awareness of environmental and social issues grows, financial institutions are increasingly recognizing that risks related to ESG factors can have profound implications for their investments and overall financial stability. By incorporating these factors into their risk management frameworks, institutions can identify, assess, and mitigate potential risks that could impact their operations and profitability.

Environmental Risks are among the most pressing concerns in today's investment landscape. Climate change, resource depletion, and pollution pose significant threats to various sectors, particularly those reliant on natural resources. For instance, companies in agriculture, mining, and energy face direct risks from changing weather patterns, regulatory changes aimed at reducing carbon emissions, and shifting consumer preferences toward sustainable products. By assessing these environmental risks, financial institutions can avoid investing in companies that may be disproportionately affected by climate change or that do not adhere to sustainable practices. This proactive approach not only protects their investments but also promotes a transition toward greener business practices.

Social Risks are equally critical in the context of ESG integration. These include issues related to labor practices, community impact, and customer relations. Companies that fail to uphold social standards, such as fair labor practices or community engagement, risk facing backlash from consumers and activists. Additionally, social inequality can lead to instability and unrest in the regions where these companies operate. By evaluating social risks, financial institutions can invest in companies that demonstrate strong social responsibility, ensuring a more stable and equitable investment landscape. This approach fosters resilience not only for their portfolios but also for the communities in which they operate, contributing to a more sustainable economy.

Governance Risks encompass issues related to corporate governance structures, including board diversity, executive pay, and transparency. Poor governance practices can lead to scandals, regulatory penalties, and ultimately, financial losses. Companies with weak governance are often less equipped to navigate crises, which can significantly affect their performance. By integrating governance factors into their risk assessments, financial institutions can better identify potential red flags and invest in companies that demonstrate strong governance practices. This not only helps mitigate risks but also encourages companies to adopt more responsible and transparent practices, thereby enhancing their long-term sustainability.

Furthermore, incorporating ESG factors into risk management processes can drive long-term value creation. As stakeholders—including investors, consumers, and regulators—place increasing importance on sustainability, companies that prioritize ESG considerations are likely to perform better over time. Financial institutions that align their investments with sustainable practices can benefit from greater market stability and resilience. This shift toward sustainable investing not only addresses immediate risks but also positions institutions to thrive in a rapidly evolving economic landscape.

Integrating ESG factors into risk management is essential for enhancing the resilience of financial portfolios. By proactively identifying and addressing environmental, social, and governance risks, financial institutions can safeguard their investments and contribute to a more sustainable future. This holistic approach not only mitigates potential long-term risks but also fosters responsible business practices that benefit society and the environment, ultimately leading to a more resilient and equitable global economy.

9.5.2.3 Financial products and services

Financial Products and Services that are aligned with sustainability goals are becoming increasingly important as investors and consumers prioritize environmental, social, and governance (ESG) criteria in their decision-making processes. By developing and offering a range of tailored financial products, institutions can facilitate the transition to a sustainable economy while also providing investment opportunities that generate financial returns. This dual focus on sustainability and profitability is crucial for driving long-term change and addressing pressing global challenges such as climate change, social inequality, and resource depletion.

Green Bonds are one of the most prominent examples of financial products designed specifically for sustainability. These fixed-income securities are issued to raise funds exclusively for projects that have positive environmental impacts, such as renewable energy installations, energy efficiency upgrades, and sustainable water management initiatives. Investors in green bonds not only contribute to financing projects that help combat climate change but also enjoy the stability and returns typically associated with fixed-income investments. The growing popularity of green bonds reflects a broader trend among investors who are increasingly seeking opportunities to align their portfolios with their values.

Social Bonds represent another important financial product aimed at addressing social challenges. These bonds fund projects that yield positive social outcomes, such as affordable housing, education, healthcare, and job creation for marginalized communities. By investing in social bonds, investors can support initiatives that improve quality of life while also earning a financial return. This dual objective aligns with the growing recognition that social equity is integral to sustainable development, and it allows investors to play a direct role in promoting social justice and community well-being.

Sustainable Investment Funds are another category of financial products that prioritize ESG criteria in their investment strategies. These funds focus on companies with strong sustainability practices, often excluding those involved in harmful activities like fossil fuels or tobacco. By investing in sustainable funds, investors can gain exposure to companies that are committed to responsible practices while also potentially achieving competitive financial returns. The increasing demand for sustainable investment funds underscores a significant shift in investor preferences, as individuals and institutions recognize the importance of sustainability in driving long-term value creation.

Green Loans and other ESG-linked financial products are also gaining traction as banks and financial institutions respond to the demand for sustainable financing options. Green loans are typically offered at preferential rates to projects that meet specific sustainability criteria, such as energy-efficient renovations or sustainable agricultural practices. Similarly, ESG-linked loans may have interest rates tied to the borrower's performance on certain sustainability metrics, incentivizing companies to improve their environmental and social practices. These innovative financial products create a direct link between sustainability performance and financial outcomes, encouraging responsible behavior while offering competitive financing options.

The development of these financial products and services not only helps channel capital toward sustainable development but also fosters a culture of responsible investing. As more investors seek to align their financial decisions with their values, the demand for sustainable financial products will continue to grow. This trend not only benefits individual investors but also creates significant opportunities for companies that prioritize sustainability, ultimately driving broader societal change.

The creation and promotion of financial products and services that align with sustainability goals are essential for facilitating the transition to a greener economy. By offering green and social bonds, sustainable investment funds, green loans, and ESG-linked financial products, financial institutions can empower investors to support sustainable development while achieving financial returns. This approach not only addresses pressing environmental and social challenges but also contributes to the long-term viability of the financial sector itself, as sustainability becomes an integral component of risk assessment and investment strategy.

9.5.2.4 Stakeholder engagement

Stakeholder Engagement is a fundamental aspect of promoting sustainable finance practices, as it fosters collaboration among various parties with vested interests in the financial ecosystem. Engaging stakeholders—including investors, companies, regulators, and civil society—ensures that diverse perspectives and expertise are integrated into the development and implementation of sustainable finance initiatives. This collaborative approach not only helps to align interests but also promotes transparency, accountability, and shared responsibility for achieving sustainability goals.

Investors play a critical role in stakeholder engagement, as they increasingly prioritize environmental, social, and governance (ESG) factors in their decision-making processes. By actively engaging with investors, financial institutions can better understand their expectations and requirements regarding sustainability practices. This dialogue enables institutions to tailor their financial products and services to meet investor demand, thereby attracting more capital for sustainable projects. Moreover, investors often share valuable insights on best practices and emerging trends, which can further enhance the effectiveness of sustainable finance strategies.

Companies, too, are key stakeholders in this engagement process. Sustainable finance initiatives can drive positive change within organizations by encouraging them to adopt responsible practices and improve their ESG performance. Engaging with companies through workshops, forums, and partnerships enables financial institutions to educate them about the benefits of sustainability, such as enhanced reputation, cost savings, and access to new markets. Collaborative efforts can lead to innovative solutions that address both financial and sustainability challenges, ultimately benefiting both companies and their investors.

Regulators also play a significant role in stakeholder engagement, as they set the legal and regulatory frameworks that govern financial markets. Engaging with regulators ensures that sustainable finance practices are aligned with policy objectives and comply with relevant laws. This dialogue can help shape regulations that incentivize sustainable investments, such as carbon pricing, disclosure requirements, and reporting standards. By collaborating with regulators, stakeholders can work together to create an enabling environment for sustainable finance that supports economic growth while addressing pressing social and environmental issues.

Furthermore, civil society organizations, including non-profits, advocacy groups, and community organizations, provide essential perspectives on the social and environmental impacts of financial practices. Engaging with these stakeholders helps financial institutions understand the broader implications of their decisions, including issues of equity, justice, and community well-being. Collaborative initiatives with civil society can also promote transparency and accountability, ensuring that sustainable finance practices truly benefit the communities they are intended to serve.

Stakeholder engagement is critical for advancing sustainable finance practices and achieving broader sustainability goals. By fostering collaboration and dialogue among investors, companies, regulators, and civil society, financial institutions can align interests, share best practices, and drive collective action. This inclusive approach not only enhances the effectiveness of sustainable finance initiatives but also contributes to the creation of a more resilient and equitable financial system. As the demand for sustainable finance continues to grow, the importance of stakeholder engagement will only increase, shaping the future of responsible investing and economic development.

9.5.2.5 Transparency and reporting

Transparency and Reporting are vital components in the realm of sustainable finance, as they enable companies and financial institutions to disclose their environmental, social, and governance (ESG) performance and impact effectively. This transparency fosters trust among stakeholders, including investors, customers, regulators, and the general public, who increasingly seek reliable information about a company's sustainability practices and overall impact on society and the environment. By ensuring that ESG policies, practices, and outcomes are clearly reported, organizations can facilitate informed decision-making among stakeholders, thereby enhancing accountability and promoting responsible investment practices.

The importance of transparent reporting lies in its ability to bridge the information gap between companies and their stakeholders. Stakeholders, especially investors, require accurate and comprehensive data to assess the sustainability risks and opportunities associated with their investments. Inadequate or misleading information can lead to misinformed decisions, which can harm both financial performance and societal well-being. Therefore, companies must prioritize transparency in their ESG disclosures, making it easier for stakeholders to evaluate their commitment to sustainability and assess their overall impact.

To standardize and enhance the quality of ESG reporting, several frameworks have been developed, including the Global Reporting Initiative (GRI) and the Sustainability Accounting Standards Board (SASB). The GRI provides a comprehensive set of guidelines for organizations to disclose their sustainability impacts across various dimensions, including environmental stewardship, social responsibility, and governance practices. By following GRI standards, companies can ensure that their reports address the most relevant ESG issues and provide a holistic view of their sustainability performance.

On the other hand, SASB focuses on industry-specific sustainability metrics that are financially material, helping companies identify and report on the ESG factors that are most likely to influence their financial performance. This targeted approach enables investors to better compare companies within the same industry, facilitating informed investment decisions based on relevant and material sustainability issues. By adopting these reporting frameworks, companies can enhance the credibility and comparability of their ESG disclosures, making it easier for stakeholders to assess their sustainability practices.

The integration of technology into ESG reporting is increasingly becoming essential. Advances in data analytics, artificial intelligence, and blockchain can enhance the accuracy and reliability of ESG data, allowing companies to track their performance more effectively and share that information transparently with stakeholders. By leveraging these technological tools, organizations can streamline their reporting processes, ensuring that they provide timely and relevant updates on their sustainability initiatives and outcomes.

Transparency in ESG reporting also serves as a tool for driving improvement within organizations. When companies publicly disclose their ESG performance, they create accountability for their commitments and actions. Stakeholders can monitor progress, provide feedback, and hold organizations accountable for their sustainability goals. This dynamic encourages companies to continuously improve their practices, ultimately leading to better environmental and social outcomes.

Transparency and reporting are critical elements of sustainable finance, as they empower stakeholders to make informed decisions based on reliable ESG information. By adopting standardized reporting frameworks such as the GRI and SASB, companies can enhance the quality and comparability of their disclosures, fostering trust and accountability. As the demand for transparency in sustainability practices continues to grow, organizations must prioritize effective ESG reporting to demonstrate their commitment to responsible business practices and contribute positively to society and the environment.

9.5.2.6 Policy and regulatory support

Policy and Regulatory Support is a crucial element in advancing sustainable finance, as it lays the groundwork for creating an enabling environment that encourages responsible investment practices and fosters a transition toward a more sustainable economy. Governments and regulators play a pivotal role in shaping the landscape of sustainable finance through the formulation and implementation of supportive policies and regulations. These initiatives not only provide the necessary framework for sustainable investments but also signal to the market the importance of integrating environmental, social, and governance (ESG) considerations into financial decision-making.

One key aspect of policy support is the introduction of incentives for green investments. Financial incentives, such as tax breaks, subsidies, or grants, can significantly lower the barriers for investors looking to support sustainable projects. For instance, governments may provide tax credits for companies investing in renewable energy technologies or energy-efficient infrastructure. By reducing the financial burden associated with these investments, policymakers can stimulate greater interest and participation in the green economy, ultimately leading to increased capital flow into sustainable initiatives.

Moreover, mandatory ESG disclosures are instrumental in ensuring transparency and accountability within the financial system. By requiring companies and financial institutions to disclose their ESG performance and practices, regulators can provide investors with the information they need to make informed decisions. This transparency enhances the overall credibility of the market, as stakeholders can assess the sustainability practices of companies before making investment choices. Mandatory reporting also encourages organizations to adopt better ESG practices, as they become aware that their performance will be scrutinized by investors and the public.

In addition to incentives and disclosure requirements, the establishment of regulatory frameworks that promote responsible financial practices is essential. These frameworks can include guidelines for integrating ESG factors into investment analysis, risk management, and corporate governance. For example, the implementation of sustainable finance regulations can encourage banks and financial institutions to assess the environmental and social impacts of their lending practices. By embedding sustainability considerations into the core of financial operations, regulators can drive systemic change within the financial sector, pushing it toward a more sustainable trajectory.

Another critical aspect of policy and regulatory support is fostering collaboration between stakeholders. Governments can facilitate dialogue among public institutions, private sector actors, and civil society to align interests and promote best practices in sustainable finance. This collaborative approach allows for the sharing of knowledge, expertise, and resources, leading to more effective and innovative solutions for addressing sustainability challenges. By engaging various stakeholders in the policymaking process, governments can create inclusive policies that reflect diverse perspectives and needs, ultimately enhancing the effectiveness of sustainable finance initiatives.

International cooperation plays a significant role in establishing a cohesive global framework for sustainable finance. Many sustainability challenges, such as climate change and biodiversity loss, transcend national borders, necessitating collaborative efforts across countries. International agreements and initiatives, such as the Paris Agreement and the Sustainable Development Goals (SDGs), provide platforms for countries to commit to sustainable finance practices and share best practices. By aligning national policies with global sustainability goals, governments can strengthen their efforts and create a more supportive environment for sustainable finance.

Policy and regulatory support are fundamental to promoting sustainable finance and creating an environment conducive to responsible investment practices. Governments can drive the transition toward a sustainable economy by implementing incentives for green investments, mandating ESG disclosures, and establishing regulatory frameworks. Collaborative efforts among stakeholders and international cooperation further enhance the effectiveness of these initiatives, ensuring that sustainability remains a core focus of financial decision-making. A robust policy landscape can empower investors, companies, and communities to pursue sustainable development goals and foster a greener, more equitable future.

9.5.3 Examples of sustainable finance in action

Examples of sustainable finance in action include issuing green bonds to fund renewable energy projects, such as solar farms and wind turbines, which help reduce carbon emissions and create jobs. Social bonds finance initiatives that address societal challenges, such as affordable housing and education, promoting social equity. Additionally, sustainable investment funds prioritize companies with strong environmental, social, and governance (ESG) practices, supporting responsible business models. Furthermore, financial institutions offer green loans with favorable terms for energy-efficient upgrades, demonstrating how financial products can align with sustainability goals, driving both economic growth and positive environmental impact.

The Green Bond Market: The green bond market has grown significantly, with issuers, including governments, municipalities, and corporations raising funds for climate-friendly projects. For example, the European Investment Bank has issued green bonds to finance European renewable energy projects (Banga, 2019).

Social Impact Bonds: Social impact bonds (SIBs) are innovative financing mechanisms that fund social programs with the promise of financial returns if the programs achieve predetermined outcomes. For instance, the UK has used SIBs to fund initiatives to reduce homelessness and improve employment outcomes for disadvantaged groups (Fraser et al., 2018).

Sustainable Investment Funds: Asset management firms such as BlackRock and Vanguard have launched sustainable investment funds that incorporate ESG criteria into their investment strategies. These funds allow investors to align their investments with their values while seeking financial returns (Mervelskemper et al., 2014).

9.5.4 The benefits of sustainable finance

Sustainable finance offers numerous benefits that contribute to environmental, social, and economic well-being. By directing capital toward green projects and initiatives, it helps reduce carbon emissions, promote renewable energy, and conserve natural resources, fostering a healthier planet. Socially, sustainable finance enhances equity by improving access to essential services and addressing societal challenges, particularly for marginalized communities. Economically, it drives innovation, creates green jobs, and strengthens resilience against long-term risks, such as climate change. Additionally, companies and financial institutions that adopt sustainable practices enhance their reputation, attracting investors and customers who value responsible and ethical practices, ultimately leading to a more sustainable future.

9.5.4.1 Environmental benefits

Transitioning to sustainable finance yields significant environmental benefits by reducing carbon emissions, promoting renewable energy, and conserving natural resources. Targeted investments in green projects, such as solar and wind energy, not only mitigate the impacts of climate change but also help protect ecosystems and biodiversity. By funding initiatives that focus on sustainability, financial institutions play a critical role in fostering a greener future.

9.5.4.2 Social benefits

Sustainable finance also brings about important social benefits. Investments in social impact projects enhance social equity and improve access to essential services, such as healthcare and education, especially for marginalized communities. By addressing societal challenges through targeted funding, sustainable finance helps build stronger, more resilient societies, ensuring that everyone has the opportunity to thrive.

9.5.4.3 Economic benefits

Economically, sustainable finance contributes to job creation and innovation. By fostering green jobs and promoting sustainable technologies, it enhances economic resilience and addresses long-term risks associated with climate change and resource depletion. This approach not only stimulates economic growth but also ensures that economies can adapt to changing conditions sustainably.

9.5.4.4 Reputational benefits

Lastly, adopting sustainable finance practices enhances the reputation and credibility of financial institutions and companies. Organizations that commit to sustainability attract investors and customers who prioritize environmentally and socially responsible practices. This positive reputation can lead to increased customer loyalty and investment, further reinforcing the benefits of sustainable finance.

Sustainable finance is an approach that integrates environmental, social, and governance criteria into financial services to support sustainable development. By investing in projects and businesses prioritizing sustainability and ethical practices, sustainable finance drives the transition to a greener and more equitable economy. Through green bonds, social bonds, sustainable investment funds, impact investing, ESG integration, and sustainable banking, sustainable finance channels capital towards initiatives that deliver long-term societal benefits.

9.6 Job Creation

Promote the creation of green jobs in renewable energy, conservation, and sustainable industries. Job creation in the context of a green economy refers to developing employment opportunities in sectors that support environmental sustainability and social well-being. This involves fostering jobs in renewable energy, conservation, and sustainable industries, essential for achieving long-term ecological and economic goals. Green jobs focus on reducing environmental impacts, conserving natural resources, and enhancing overall sustainability (Doan and Nguyen, 2023; Lay and Tafese, 2020; Norbu et al., 2021).

The significance of creating green jobs extends beyond environmental benefits to include economic and social impacts. Green jobs can drive economic growth, support community development, and address unemployment, especially in regions affected by traditional industry declines. By prioritizing green job creation, societies can ensure a transition to a more sustainable economy while providing meaningful employment opportunities.

9.6.1 Key areas for green job creation

Key areas for green job creation include renewable energy, energy efficiency, and sustainable agriculture. Jobs in renewable energy, such as wind, solar, and hydropower, are expanding rapidly as countries transition to cleaner energy sources. Energy efficiency initiatives in industries like construction, manufacturing, and transportation also create jobs in retrofitting buildings, optimizing energy use, and developing cleaner technologies. Sustainable agriculture, which focuses on practices like organic farming, agroforestry, and regenerative techniques, offers employment opportunities while preserving ecosystems. Additionally, sectors like waste management, water conservation, and sustainable transportation play a crucial role in creating jobs that support environmental sustainability.

9.6.1.1 Renewable energy

This sector includes jobs related to the development, installation, and maintenance of renewable energy sources such as solar, wind, hydro, and geothermal power. Positions in this field range from solar panel installers and wind turbine technicians to energy system engineers and project managers. The renewable energy sector is a significant driver of green job growth due to its expanding role in the global energy transition.

9.6.1.2 Energy efficiency

Energy efficiency jobs involve improving energy use in buildings, transportation, and industry. These include roles such as energy auditors, building retrofitting specialists, and energy-efficient appliance designers. By optimizing energy consumption and reducing waste, these jobs contribute to lower carbon emissions and reduced energy costs.

9.6.1.3 Conservation and environmental management

Careers in this area focus on protecting and restoring natural ecosystems, managing natural resources, and mitigating environmental degradation. Positions include conservation scientists, environmental engineers, wildlife biologists, and land restoration experts. These roles are crucial for maintaining biodiversity, managing protected areas, and addressing environmental challenges.

9.6.1.4 Sustainable agriculture and forestry

This sector encompasses sustainable farming practices, agroecology, and forest management jobs. Roles include organic farmers, sustainable crop consultants, forest rangers, and agroforestry specialists. Sustainable agriculture and forestry practices help preserve soil health, reduce pesticide use, and manage forests responsibly.

9.6.1.5 Green transportation

Jobs in green transportation focus on developing and implementing sustainable transportation solutions. This includes roles in electric vehicle manufacturing, public transit planning, and bike-sharing program management. Promoting cleaner and more efficient transportation options, these jobs help reduce greenhouse gas emissions and improve urban mobility.

9.6.1.6 Circular economy

The circular economy involves creating jobs related to recycling, waste management, and product design for longevity and reuse. Positions in this field include recycling plant operators, waste reduction consultants, and product designers focused on circular design principles. These jobs help reduce waste and extend the lifecycle of products.

9.6.2 The process of green job creation

The process of green job creation begins with investments in sustainable industries and technologies, such as renewable energy, energy efficiency, and sustainable agriculture. Governments and businesses play a key role by implementing supportive policies, incentives, and training programs to build the necessary workforce. As demand for clean energy, resource efficiency, and environmentally responsible practices increases, new opportunities emerge in areas like solar and wind energy installation, green building construction, and sustainable farming. Additionally, re-skilling and up-skilling workers for green technologies are essential to ensure the workforce adapts to the evolving needs of a sustainable economy.

9.6.2.1 Policy and regulation

Governments play a crucial role in fostering green job creation through policies and regulations that support sustainability. This includes implementing incentives for renewable energy projects, setting energy efficiency standards, and funding conservation programs. Supportive policies can attract investment, stimulate job growth, and create a favorable environment for green industries.

9.6.2.2 Education and training

Providing education and training programs is essential for equipping individuals with the skills needed for green jobs. This includes developing curricula in environmental sciences, renewable energy technologies, and sustainable practices. Vocational training, apprenticeships, and higher education programs can prepare the workforce for emerging green sectors.

9.6.2.3 Public and private sector investment

Public and private sector investments are necessary to drive green job creation. Governments can invest in infrastructure projects, research and development, and job training programs. Private companies can contribute by investing in sustainable technologies, green innovations, and environmentally responsible practices. Collaborative efforts between sectors can enhance job creation and economic growth.

9.6.2.4 Supporting innovation

Encouraging innovation in green technologies and sustainable practices can create new job opportunities. This includes funding research and development, supporting startups in green industries, and promoting technological advancements. Innovation can drive the growth of emerging sectors and create high-quality jobs.

9.6.2.5 Community engagement

Involving communities in green job initiatives can enhance local development and ensure job creation meets diverse populations' needs. Community-based projects, regional partnerships, and grassroots initiatives can support job creation in urban agriculture, local conservation efforts, and sustainable tourism.

9.6.3 Examples of successful green job creation

Successful examples of green job creation can be seen in countries that have invested heavily in renewable energy and sustainable industries. For instance,

Germany's Energiewende (energy transition) initiative has generated thousands of jobs in the wind and solar energy sectors. Denmark has also created numerous green jobs through its leadership in wind energy production. In the U.S., states like California have seen significant job growth in the clean energy industry, particularly in solar panel installation and energy efficiency upgrades. Countries like Brazil have also promoted green jobs through sustainable agriculture and forest conservation initiatives.

Germany's Energiewende: Germany's transition to renewable energy has created numerous jobs in solar and wind energy sectors. The country's focus on green technology and energy efficiency has also led to job growth in energy auditing, retrofitting, and sustainable construction (Quitzow et al., 2016).

The U.S. Green New Deal: Various states in the U.S. have implemented green job programs as part of broader climate and economic policies. For example, California's investments in clean energy have generated jobs in solar energy installation, electric vehicle manufacturing, and energy efficiency retrofits (Galvin and Healy, 2020).

China's Green Development: China has become a global leader in renewable energy production, particularly solar and wind power. The expansion of these industries has resulted in significant job creation in manufacturing, installation, and maintenance (Harlan, 2021).

9.6.4 The benefits of green job creation

Green job creation offers multiple benefits, contributing to both environmental sustainability and economic growth. It helps reduce carbon emissions and environmental degradation by focusing on industries like renewable energy, sustainable agriculture, and conservation. Green jobs promote economic resilience by providing employment in emerging sectors that address climate change, boosting local economies and fostering innovation. Moreover, these jobs often improve social equity by creating opportunities for marginalized communities, contributing to poverty reduction, and enhancing overall well-being. By aligning economic activities with sustainability goals, green job creation supports long-term environmental, economic, and social benefits.

Environmental Benefits: Reducing carbon emissions, conserving natural resources, and improving environmental quality through sustainable practices.

Economic Benefits: Stimulating economic growth, diversifying local economies, and creating resilient job opportunities.

Social Benefits: Enhancing community well-being, providing meaningful employment, and addressing social inequalities through inclusive job creation.

Green job creation fosters employment opportunities in renewable energy, conservation, sustainable industries, and other areas that support environmental sustainability and social well-being ((Doan and Nguyen, 2023; Lay and Tafese, 2020; Norbu et al., 2021). By implementing supportive policies, investing in education and training, and encouraging innovation, societies can drive job growth while advancing toward a more sustainable and resilient economy.

10

Health and Well-being

Health and well-being management encompasses a broad spectrum of issues that impact individuals and communities, from preventing and treating diseases to promoting mental health and overall quality of life. Effective management in this area is crucial for improving public health outcomes and ensuring people live healthier, more fulfilling lives. However, several significant problems complicate the management of health and well-being (Topp et al., 2015; Erhart et al., 2009; Gilmour, 2012).

One major issue is the disparity in access to healthcare services. Socioeconomic factors, geographic location, and systemic inequalities can lead to uneven access to medical care. In many regions, particularly in low-income and rural areas, there is a shortage of healthcare facilities and professionals, which limits people's ability to receive timely and adequate care. This disparity can result in poorer health outcomes and a lower quality of life for those affected. Addressing this issue requires targeted policies to increase healthcare accessibility, improve infrastructure, and distribute healthcare resources more equitably.

Another significant problem is the rising prevalence of chronic diseases. Conditions such as diabetes, heart disease, and obesity are becoming more common worldwide, driven by factors like sedentary lifestyles, poor dietary habits, and environmental influences. Managing chronic diseases requires ongoing medical attention and lifestyle modifications, which can be challenging for individuals and healthcare systems alike. Effective management strategies must include prevention efforts, such as promoting healthy lifestyles, early detection, and comprehensive care plans to manage these conditions and improve patients' quality of life.

Mental health issues also pose a critical challenge. Mental health disorders, including depression, anxiety, and stress-related conditions, are increasingly recognized as significant contributors to the overall burden of disease. Despite growing awareness, mental health services are often underfunded and stigmatized, leading to inadequate support for those in need. Addressing mental health requires integrating mental health services into primary care, reducing stigma, and increasing access to counseling and psychological support.

The integration of health services represents another challenge. Many healthcare systems operate in fragmented ways, with poor coordination between primary care, specialty care, and community services. This fragmentation can lead to inefficiencies, gaps in care, and a lack of continuity for patients. Developing integrated care models that streamline services and improve communication between different levels of care is essential for enhancing patient outcomes and ensuring comprehensive health and well-being management.

The rising cost of healthcare is also a significant concern. As medical technology advances and the demand for healthcare services increases, costs can become prohibitive for individuals and governments alike. High healthcare costs can strain patients financially, particularly those without adequate insurance coverage. Addressing this issue involves exploring cost-effective treatment options, improving healthcare efficiency, and developing policies to ensure that healthcare remains affordable and accessible.

Finally, health promotion and disease prevention efforts are often insufficient. Many health issues could be mitigated through proactive measures such as vaccination programs, health education, and preventive screenings. However, these efforts can be underfunded or poorly implemented, resulting in missed opportunities for reducing the incidence of preventable diseases. Strengthening preventive health strategies and ensuring they reach all population segments is crucial for improving overall health outcomes.

Managing health and well-being involves addressing multiple complex problems, including disparities in access to care, the rising prevalence of chronic diseases, mental health issues, fragmented service integration, high healthcare costs, and insufficient preventive efforts. Effective management requires a comprehensive approach that includes improving accessibility, integrating services, addressing mental health needs, and promoting preventive measures (Topp et al., 2015; Erhart et al., 2009; Gilmour, 2012). By tackling these challenges, societies can enhance health outcomes and improve the quality of life for individuals and communities.

10.1 Healthy Lifestyles

Promote healthy lifestyles for individuals and the environment, such as plant-based diets and active transportation. Healthy lifestyles are integral to improving individual well-being and supporting environmental sustainability. Promoting healthy behaviors can significantly benefit both personal health and the environment. Key components of healthy lifestyles include adopting plant-based diets and engaging in active transportation. These practices not only contribute to better health outcomes but also help reduce environmental impacts (Martín-María et al., 2020; Fromel et al., 2017; Hautekiet et al., 2022).

Adopting plant-based diets is one of the most effective ways to promote personal health and environmental sustainability. These diets focus on fruits, vegetables, grains, nuts, and legumes while minimizing or eliminating animal products. They offer numerous health benefits and are often associated with lower risks of chronic diseases such as heart disease, hypertension, diabetes, and certain types of cancer. Plant-based diets are rich in essential nutrients, fiber, and antioxidants, contributing to overall health and vitality.

From an environmental perspective, plant-based diets can significantly reduce the ecological footprint associated with food production. Plant-based foods generally require fewer natural resources, such as water and land, than animal-based foods. Additionally, plant-based diets contribute to lower greenhouse gas emissions, as livestock farming is a major source of methane and other greenhouse gases. Individuals can help mitigate climate change and promote more sustainable food systems by shifting towards plant-based eating.

Active transportation is another crucial aspect of a healthy lifestyle that benefits individuals and the environment. Active transportation refers to modes of travel that require physical activity, such as walking, cycling, and using non-motorized scooters. Active transportation helps improve cardiovascular health, maintain a healthy weight, and enhance overall fitness. It also reduces the risk of chronic diseases linked to sedentary lifestyles.

The environmental benefits of active transportation are equally significant. Individuals can reduce their carbon footprint and decrease air pollution by walking or cycling instead of driving. Transportation is an important contributor to greenhouse gas emissions, and reducing reliance on fossil-fuel-powered vehicles helps mitigate climate change. Active transportation also contributes to reduced traffic congestion and lower noise pollution, leading to more pleasant and livable urban environments.

Promoting healthy lifestyles involves creating supportive environments and providing resources to make these practices more accessible and appealing. This includes designing cities with safe and convenient walking and cycling infrastructure, such as bike lanes, pedestrian crossings, and green spaces. Public policies and urban planning should prioritize these elements to encourage active transportation and make it a viable option for more people.

Educational campaigns and community programs can also play a crucial role in promoting plant-based diets and active transportation. Information about these practices' health and environmental benefits and practical tips and resources for making lifestyle changes can help individuals make informed choices and adopt healthier behaviors.

Promoting healthy lifestyles through plant-based diets and active transportation substantially benefits individual health and environmental sustainability. Plant-based diets improve health outcomes and reduce environmental impact, while active transportation supports physical fitness and lowers carbon emissions (Martín-María et al., 2020; Fromel et al., 2017; Hautekiet et al., 2022). By creating supportive environments, implementing public policies, and raising awareness, societies can encourage adopting these practices and enhance overall well-being.

10.2 Mental Health

Address the mental health impacts of environmental degradation and climate change. Environmental degradation and climate change profoundly influence mental health, impacting individuals' emotional and psychological well-being in various ways. Addressing these impacts is crucial for fostering resilience and promoting overall mental health in the face of environmental challenges (Mcdaid and Park, 2011; Otto et al., 2021; Kelly et al., 2011).

One significant mental health impact of environmental degradation is the phenomenon known as "eco-anxiety." Eco-anxiety refers to the feelings of distress, worry, or helplessness related to environmental issues such as climate change, pollution, and biodiversity loss. Individuals experiencing eco-anxiety may struggle with constant fears about the future, concerns for the well-being of future generations, and a sense of powerlessness in the face of overwhelming environmental problems. This form of anxiety can lead to symptoms such as chronic stress, depression, and insomnia.

Another mental health concern is the psychological impact of extreme weather events and natural disasters. Events such as hurricanes, floods, wildfires, and droughts can have devastating effects on mental health. People affected by these disasters often experience trauma, post-traumatic stress disorder (PTSD), and other stress-related conditions. The loss of homes, livelihoods, and loved ones, combined with the disruption of daily life, can lead to long-term psychological impacts. Recovery from such events requires physical rebuilding, mental health support, and community resilience efforts.

The gradual impacts of climate change can also contribute to mental health issues. For instance, prolonged heatwaves and poor air quality can exacerbate pre-existing mental health conditions and contribute to feelings of frustration and helplessness. Additionally, the stress of adapting to changing environmental conditions, such as increased food insecurity or displacement from rising sea levels, can strain mental health and well-being.

10.2.1 Addressing mental health impacts involves several key strategies

Addressing mental health impacts requires a multi-faceted approach that includes increasing access to mental health services, promoting early intervention, and reducing stigma. Ensuring that mental health care is integrated into primary health systems can improve accessibility for needy individuals. Public awareness campaigns can help reduce the stigma associated with mental health issues, encouraging people to seek help. Promoting self-care, resilience, and community-based support networks are important for prevention and recovery. Policies that prioritize mental well-being in schools, workplaces, and communities also play a critical role in addressing the broader impacts on mental health.

10.2.1.1 Integrating mental health support into environmental policies

Recognizing the intersection of environmental issues and mental health within policy frameworks is crucial. This includes incorporating mental health considerations into climate adaptation and resilience planning and ensuring that mental health services are available and accessible to those affected by environmental issues.

10.2.1.2 Providing access to mental health services

Expanding access to mental health services, including counseling and therapy, is essential for addressing eco-anxiety and trauma related to environmental degradation. Providing support in the aftermath of natural disasters, offering crisis intervention, and promoting mental health awareness can help individuals cope with the psychological impacts of environmental stressors.

10.2.1.3 Promoting community resilience

Building community resilience through support networks, social cohesion, and disaster preparedness can mitigate the mental health impacts of environmental events. Community-based initiatives, such as support groups and local preparedness programs, can offer emotional support and practical assistance during and after environmental crises.

10.2.1.4 Encouraging eco-therapy and nature-based interventions

Engaging with nature has been shown to have therapeutic benefits. Eco-therapy, which involves spending time in natural environments and connecting with nature, can help alleviate stress, anxiety, and depression. Promoting access to green spaces and nature-based interventions can support mental well-being and foster a sense of environmental connection.

10.2.1.5 Raising awareness and reducing stigma

It is crucial to raise awareness about the mental health impacts of environmental issues and reduce stigma around seeking help. Educational campaigns can inform the public about the psychological effects of environmental stressors and encourage individuals to seek support when needed.

10.2.1.6 Supporting research on environmental and mental health connections

Investing in research to better understand the relationship between environmental degradation, climate change, and mental health is crucial for developing effective interventions. Research can provide insights into the mechanisms of eco-anxiety, the psychological impacts of different environmental stressors, and practical strategies for support.

Addressing the mental health impacts of environmental degradation and climate change involves recognizing and responding to the psychological effects of eco-anxiety, trauma from natural disasters, and the gradual stress of climate change. Integrating mental health into environmental policies, expanding access to services, promoting community resilience, and encouraging nature-based interventions are essential for supporting mental well-being in environmental challenges (Mcdaid and Park, 2011; Otto et al., 2021; Kelly et al. 2011). By addressing these issues, societies can foster resilience, improve mental health outcomes, and enhance overall well-being in an era of ecological uncertainty.

11

Ecosystem Services

Recognize and preserve the natural services provided by ecosystems, such as clean air, water, and pollination. Ecosystem services are the essential benefits humans derive from natural ecosystems, which are crucial for sustaining life and supporting human well-being. These services encompass various ecosystem functions, including regulating air and water quality, nutrient cycling, and pollination. Recognizing and preserving these services is vital for maintaining ecological balance and ensuring the continued availability of resources upon which human societies depend (Harrison et al., 2014; Elmqvist et al., 2015; Carpenter et al., 2006; Martin-Ortega et al., 2015).

Major ecosystem services

1. Provisioning Services:

   Food Production: Providing crops, livestock, fisheries, and aquaculture.
   Water Supply: Freshwater for drinking, irrigation, and sanitation.
   Raw Materials: Timber, fiber, and other materials for construction and fuel.
   Medicinal Resources: Plants and natural compounds used in pharmaceuticals.

2. Regulating Services:

   Climate Regulation: Carbon sequestration, temperature and precipitation control.
   Water Purification: Wetlands and forests filter pollutants and improve water quality.
   Pollination: Bees, birds, and insects pollinate crops and wild plants.
   Erosion Control: Vegetation preventing soil erosion and landslides.
   Pest and Disease Control: Natural predators regulating pests and diseases.

3. Supporting Services:

   Nutrient Cycling: Decomposition and nutrient recycling essential for soil fertility.
   Soil Formation: Weathering of rocks and organic matter contributing to soil formation.
   Photosynthesis: Plants convert sunlight into energy, forming the basis of the food chain.

4. Cultural Services:

   Recreational Opportunities: Nature-based tourism, hiking, and outdoor activities.
   Aesthetic Value: Scenic landscapes and natural beauty.
   Spiritual and Religious Significance: Sacred sites and spiritual connections with nature.
   Educational and Research Opportunities: Natural areas for study and learning.

Source: Bhandari, 2019.

The most fundamental ecosystem service is the provision of clean air. Forests, wetlands, and other natural environments are crucial in filtering pollutants from the air and maintaining atmospheric quality. Trees and vegetation absorb carbon dioxide and release oxygen through photosynthesis, contributing to a stable climate and reducing the impacts of air pollution. Urban green spaces, such as parks and urban forests, also help mitigate air pollution by trapping dust and absorbing pollutants, enhancing public health and quality of life.

Clean water is another critical ecosystem service provided by natural systems. Ecosystems such as wetlands, forests, and riparian zones (areas adjacent to rivers and streams) play a vital role in water purification. Wetlands act as natural filters, trapping sediments and absorbing excess nutrients from runoff before they enter water bodies. Forests help maintain watershed health by regulating water flow, preventing soil erosion, and reducing the risk of floods. Protecting these ecosystems ensures that water resources remain clean and available for drinking, agriculture, and other uses.

Pollination is an essential ecosystem service provided by various organisms, including bees, butterflies, birds, and bats. Pollinators are crucial for reproducing many flowering plants and crops, directly affecting food production and biodiversity. Approximately one-third of the world's food crops rely on animal pollinators, making this service critical for food security and agricultural sustainability. Conserving habitats that support pollinators, such as wildflower meadows and diverse agricultural landscapes, is essential for maintaining healthy ecosystems and ensuring crop yields.

11.1 Preserving Ecosystem Services Involves Several Key Strategies

Preserving ecosystem services is critical for maintaining the planet's health and human well-being, and it involves several key strategies. First, protecting natural habitats through conservation efforts and sustainable land-use practices helps maintain biodiversity, essential for ecosystem resilience. Second, restoring degraded ecosystems, such as reforestation and wetland restoration, enhances their capacity to provide vital services like carbon sequestration, water purification, and climate regulation. Third, promoting sustainable resource management practices, such as responsible farming, fishing, and forestry, ensures that ecosystems are not overexploited. Fourth, adopting nature-based solutions for addressing environmental challenges, such as using green infrastructure to manage stormwater, helps reduce human impact while benefiting ecosystems. Lastly, engaging local communities in the management and conservation of ecosystems fosters stewardship and ensures that ecosystems continue providing services that support livelihoods and broader environmental sustainability. Together, these strategies ensure the long-term viability of ecosystem services.

11.1.1 Protecting natural habitats

Protecting natural habitats is essential for preserving the ecosystem services that sustain life on Earth, including clean air, water, fertile soil, and biodiversity. Safeguarding natural areas such as forests, wetlands, grasslands, and marine ecosystems ensures the continued functioning of ecological processes that provide these critical services. One of the most effective strategies for this protection is the establishment of protected areas, such as national parks, wildlife reserves, and conservation corridors. These areas act as refuges for plant and animal species, maintaining biodiversity and the resilience of ecosystems in the face of environmental pressures like climate change.

In addition to creating protected areas, land-use planning and zoning regulations play a crucial role in preventing habitat destruction and degradation. Governments and local authorities can implement land-use policies that prioritize conservation, limiting urban sprawl, deforestation, and agricultural expansion into sensitive habitats. Integrating conservation priorities into broader development plans can help balance economic growth with the need to preserve ecosystems. For example, enforcing buffer zones around protected areas can prevent encroachment and reduce the impact of human activities on vulnerable ecosystems.

Community-based conservation efforts are increasingly recognized as an important element in habitat protection. By involving local populations in the management of natural resources and creating economic incentives for conservation—such as eco-tourism or sustainable harvesting—communities become stewards of their local environments. This not only protects habitats but also ensures the sustainability of the ecosystem services on which these communities depend for their livelihoods.

International cooperation is also key to protecting transboundary ecosystems that cross national borders. Collaborative conservation efforts in regions like the Amazon rainforest or the Great Lakes can help ensure the protection of ecosystems that provide global services, such as carbon sequestration and climate regulation. The combined efforts of local, national, and international stakeholders are therefore critical to safeguarding natural habitats and maintaining the services they provide for future generations.

11.2 The Process to Restore – Rebuilt the Disbalanced Ecosystems

It is essential to adopt a holistic approach to restoring and rebuilding imbalanced ecosystems caused by unsustainable use of nature. The process begins with assessment, identifying the extent of degradation, and understanding the ecosystem's original state. This includes mapping biodiversity loss, soil degradation, water quality, and habitat fragmentation. Restoration planning follows, involving strategies to rehabilitate natural resources, reintroduce native species, and remove invasive species that harm the ecosystem.

Conservation of remaining natural areas is crucial to prevent further degradation. Sustainable land management practices, such as reforestation, soil regeneration, and water conservation, help rebuild ecosystem services. Community involvement ensures local knowledge and participation in restoration efforts, fostering long-term commitment.

Another key element is promoting biodiversity, which involves re-establishing native flora and fauna, enhancing habitat connectivity, and protecting key species to maintain ecological balance. Monitoring and adaptive management are necessary to track progress and adjust restoration strategies as needed.

Integrating policy support for sustainable practices and environmental protection is essential to ensure long-term success. This comprehensive approach fosters a resilient and healthy ecosystem that can sustain both human and non-human life while mitigating the impacts of climate change.

11.2.1 Restoring degraded ecosystems

Restoring degraded ecosystems is a crucial step in addressing the environmental impacts of human activities that have damaged natural habitats, disrupted ecological balance, and reduced biodiversity. Ecosystems are the foundation of life on Earth, providing essential services such as clean air and water, fertile soils, and food security. However, unsustainable practices such as deforestation, overgrazing, pollution, and urban sprawl have severely degraded many ecosystems worldwide, threatening wildlife and human well-being. Restoration efforts aim to rehabilitate these damaged ecosystems, allowing them to recover their functionality, resilience, and biodiversity.

One of the most common restoration methods is reforestation, which involves planting trees in areas that have been deforested or degraded. Forests play a vital role in carbon sequestration, water cycle regulation, and habitat provision for numerous species. Reforestation not only helps restore ecosystem services but also mitigates climate change by absorbing atmospheric carbon dioxide. In many restoration projects, a focus on planting native tree species is essential, as these species are better suited to local conditions and support a wider range of biodiversity than non-native species.

Wetland reconstruction is another key component of ecosystem restoration. Wetlands are highly productive ecosystems that provide critical services, such as water filtration, flood control, and carbon storage. However, many wetlands have been drained or altered for agricultural and urban development. Restoration of these areas often involves re-establishing natural water flows, removing pollutants, and reintroducing native plant and animal species. Wetland restoration not only improves water quality and reduces flood risks but also supports species such as birds, amphibians, and fish that rely on these habitats.

The removal of invasive species is also an important aspect of restoration. Invasive species can outcompete native species, alter habitats, and disrupt ecosystem processes. By removing invasive plants and animals, restoration projects can help native species recover and re-establish the ecological balance. This is particularly important in ecosystems that have been heavily altered by human activities, as invasive species often thrive in disturbed environments. Removal efforts must be carefully managed to avoid unintended consequences and should be coupled with reintroducing native species and restoring natural processes.

In addition to these specific techniques, restoration projects aim to enhance the overall resilience of ecosystems by supporting biodiversity and ecological connectivity. Healthy ecosystems are better able to withstand disturbances such as climate change, natural disasters, and human activities. By promoting biodiversity, restoration efforts ensure that ecosystems have the diversity of species and functions needed to maintain their balance and adapt to changing conditions. Enhancing habitat connectivity—by creating wildlife corridors, restoring fragmented habitats, and protecting key ecosystems—allows species to migrate, reproduce, and maintain genetic diversity, which is essential for long-term ecosystem health.

Restoring ecosystems is about more than just repairing damage; it is about fostering the natural processes that sustain life. Through a combination of reforestation, wetland reconstruction, invasive species removal, and promoting biodiversity, ecosystem restoration projects can help rebuild ecosystems' capacity to provide essential services, support biodiversity, and contribute to climate change mitigation. These efforts also require the involvement of local communities, policy support, and long-term monitoring to ensure lasting success.

11.2.2 Implementing sustainable practices

Implementing sustainable practices in agriculture, forestry, and land management is essential for reducing the negative impacts of human activities on ecosystems and ensuring their long-term health. As human populations grow and demand for natural resources increases, it is critical to adopt methods that meet human needs and maintain the integrity of ecosystems. Sustainable practices prioritize the conservation of biodiversity, the responsible use of resources, and the minimization of harmful environmental impacts, helping to preserve the services that ecosystems provide to both humans and wildlife.

In agriculture, unsustainable practices such as monocropping, excessive use of chemical fertilizers and pesticides, and over-irrigation have contributed to soil degradation, water pollution, and loss of biodiversity. However, sustainable approaches like agroforestry offer a promising alternative. Agroforestry integrates trees and shrubs into agricultural landscapes, creating a system that enhances biodiversity, improves soil health, and increases carbon sequestration. Trees in agroforestry systems can provide shade, prevent soil erosion, and improve water retention, leading to more resilient and productive agricultural ecosystems. Additionally, agroforestry can provide farmers with additional sources of income from timber, fruits, and other tree products, making it an economically viable and ecologically sound practice.

Another important sustainable agricultural technique is conservation tillage, which involves reducing or eliminating plowing to maintain soil structure and prevent erosion. Traditional tillage disrupts soil ecosystems, leading to the loss of organic matter and soil biodiversity. Conservation tillage, on the other hand, leaves crop residues on the field, which helps retain moisture, reduce soil erosion, and increase carbon storage in the soil. This practice not only improves soil fertility but also supports a healthier soil microbiome, which is essential for plant growth and resilience. By minimizing soil disturbance, conservation tillage can also reduce greenhouse gas emissions from agricultural fields, contributing to climate change mitigation.

In forestry, sustainable management practices such as selective logging, replanting, and reducing clear-cutting help maintain forest ecosystems' biodiversity and productivity. Sustainable forestry involves harvesting trees in a way that allows the forest to regenerate naturally, ensuring that timber resources are available for future generations. Selective logging, for example, removes only certain trees while leaving the rest of the forest intact, preserving habitats for wildlife and maintaining the ecological functions of the forest. By managing forests sustainably, we can protect biodiversity, prevent soil erosion, and enhance the forest's ability to sequester carbon, thus playing a key role in climate change mitigation efforts.

Sustainable fisheries are another critical area where sustainable practices can prevent environmental degradation and promote ecosystem health. Overfishing has led to the collapse of many fish populations and disrupted marine ecosystems. Sustainable fisheries management involves setting catch limits, protecting critical habitats such as coral reefs and mangroves, and enforcing regulations to prevent illegal fishing practices. By using more selective fishing gear and reducing bycatch, sustainable fisheries help preserve marine biodiversity and ensure that fish populations can replenish themselves. Protecting marine ecosystems is vital for maintaining the balance of oceanic food webs, supporting livelihoods, and ensuring food security for millions of people.

In land management, practices such as integrated pest management (IPM) and rotational grazing can also contribute to ecosystem health. IPM reduces the need for chemical pesticides by using biological controls, crop rotation, and other techniques to manage pests in a more environmentally friendly manner. Rotational grazing involves moving livestock between pastures to prevent overgrazing, allowing grasslands to recover and maintain their biodiversity and productivity.

By adopting these sustainable practices across various sectors, we can reduce environmental degradation, protect ecosystems, and enhance their resilience. These practices not only benefit the environment but also contribute to long-term economic sustainability by ensuring that natural resources remain available for future generations. Ultimately, the widespread implementation of sustainable practices is key to fostering a more harmonious relationship between human activities and the natural world.

11.2.3 Promoting ecosystem-based management (EBM)

Promoting Ecosystem-Based Management is a holistic approach that recognizes the interconnectedness of ecosystems and the vital services they provide to both human and non-human life. Rather than focusing on a single species or resource in isolation, EBM considers the complex relationships within ecosystems, including interactions among species, habitats, and environmental factors, as well as the ways these systems support human communities. This approach is essential for maintaining ecological balance, protecting biodiversity, and ensuring the sustainable use of natural resources.

A key principle of EBM is the integration of ecological, social, and economic objectives in resource management. By recognizing that ecosystems are dynamic and interconnected, EBM seeks to manage natural resources in a way that preserves ecosystem integrity while allowing for sustainable human use. This includes maintaining healthy ecosystems that provide essential services such as clean water, air purification, pollination, and climate regulation. For example, managing a watershed through an ecosystem-based lens would involve not just regulating water extraction but also protecting forested areas, wetlands, and riparian zones that contribute to water quality and biodiversity.

One of the strengths of EBM is its emphasis on using scientific knowledge to inform decision-making. Ecological monitoring, data collection, and research help managers understand the current state of ecosystems, identify threats, and develop strategies to mitigate those threats. Science-based management ensures that actions are guided by the best available evidence, whether it involves setting sustainable fishing quotas, restoring degraded habitats, or protecting endangered species. However, EBM also recognizes the value of traditional knowledge and practices, which have often been developed over centuries by indigenous and local communities. Traditional ecological knowledge can complement scientific understanding by offering insights into ecosystem functions, species behaviors, and sustainable practices that have stood the test of time.

Another critical aspect of EBM is stakeholder engagement. Effective ecosystem management requires the active participation of various stakeholders, including local communities, government agencies, conservation organizations, and industries that rely on natural resources. Engaging stakeholders in decision-making ensures that different perspectives are considered and that management strategies are more likely to be supported and implemented effectively. When communities that depend on an ecosystem's resources, such as farmers, fishers, or indigenous groups, are involved in the process, the likelihood of long-term success increases, as they are more invested in the health and sustainability of the ecosystem.

EBM also promotes adaptive management, recognizing that ecosystems are constantly changing due to factors such as climate variability, human activities, and natural disturbances. This requires flexibility in management approaches, with regular monitoring and adjustment of strategies based on new information and changing conditions. By using feedback loops, EBM allows for continuous learning and refinement, helping resource managers respond to unexpected challenges and improve outcomes over time.

In practical terms, EBM can be applied across various sectors, including marine and coastal management, forestry, agriculture, and water resource management. In marine environments, for example, ecosystem-based fisheries management seeks to balance the needs of fishing communities with the preservation of marine biodiversity. This involves setting sustainable catch limits, protecting critical habitats, and considering the broader impacts of fishing activities on the marine ecosystem as a whole. In forest management, EBM might include managing logging practices in ways that preserve forest structure and biodiversity while also considering the impacts on water cycles, wildlife habitats, and the rights of indigenous communities.

By focusing on the whole system and considering ecological, social, and economic factors, EBM represents a sustainable approach to managing natural resources in the face of growing environmental pressures. Its emphasis on integration, collaboration, and flexibility makes it well-suited to addressing complex challenges such as climate change, habitat loss, and resource depletion, ultimately supporting both human and ecological well-being.

11.2.4 Raising awareness and education

Raising awareness and education about the importance of ecosystem services and the threats they face is a fundamental strategy for fostering support for conservation and sustainable practices. Ecosystem services, such as clean air, water purification, pollination, and climate regulation, are often taken for granted because their benefits are not always visible to the public. As a result, people may not fully appreciate the urgency of protecting these services or understand how human activities contribute to their degradation. Education and outreach can help bridge this gap by empowering individuals and organizations to take action in preserving ecosystems.

One of the most effective ways to raise awareness is through educational programs that target different audiences, from schoolchildren to policymakers. These programs can teach people about the intricate relationships within ecosystems, the services they provide, and how their health is linked to human well-being. Integrating environmental education into school curriculums is especially important, as it instills ecological awareness and responsibility in future generations. Field trips to natural reserves, hands-on environmental projects, and interactive learning experiences can foster a deeper understanding of nature and the importance of conservation. Beyond schools, educational programs for adults, including workshops and seminars, can provide opportunities to learn about sustainable practices, such as reducing waste, conserving water, and supporting biodiversity.

Outreach campaigns are another critical tool for raising awareness. By using a variety of communication channels, such as social media, public service announcements, documentaries, and informational websites, outreach campaigns can reach a broad audience and highlight the urgent need for ecosystem protection. These campaigns can focus on specific issues, such as the decline of pollinators, deforestation, or plastic pollution, while offering actionable steps that individuals and businesses can take to mitigate these problems. Storytelling can be a powerful element of outreach, as sharing success stories of conservation efforts or illustrating the impact of ecosystem degradation on local communities helps make the issue more relatable and compelling.

Community engagement initiatives also play a key role in raising awareness at the grassroots level. By involving local communities in conservation projects, such as tree planting, wetland restoration, or wildlife monitoring, people can see firsthand the positive impact of protecting and restoring ecosystems. Participatory approaches that incorporate local knowledge and values can create a sense of ownership and responsibility for environmental stewardship. Community engagement fosters not only environmental awareness but also social cohesion, as people work together towards a common goal of protecting the natural resources that sustain their community.

Collaborations between environmental organizations, government agencies, businesses, and academic institutions can amplify the reach of awareness and education efforts. Partnerships can enhance the credibility and impact of campaigns, as well as provide resources and expertise that strengthen conservation initiatives. For example, corporations can play a role by integrating sustainability principles into their operations and educating their employees and customers about the importance of ecosystem services. Universities can contribute by conducting research on ecosystems and providing scientific data that informs public policy and conservation strategies.

Public awareness and education help cultivate a deeper appreciation for ecosystems and encourage responsible behavior that benefits both nature and society. When individuals understand the value of ecosystem services and recognize the threats they face, they are more likely to support policies, practices, and initiatives that promote conservation. By raising awareness, we can inspire people to become advocates for the environment, engage in sustainable practices, and participate in efforts to restore and protect the ecosystems that are essential to all life on Earth.

11.2.5 Incorporating ecosystem services into policy and planning

Incorporating ecosystem services into policy and planning is a critical step toward ensuring that the value of nature is fully considered in decision-making processes. Ecosystem services, such as water purification, carbon sequestration, and flood regulation, are essential to human well-being and economic development. However, these services are often overlooked or undervalued in traditional policy frameworks. By integrating ecosystem services into policy and planning, governments and organizations can make more informed decisions that balance economic growth with environmental sustainability, ultimately leading to better outcomes for both people and the planet.

One of the key tools for achieving this integration is ecosystem service assessments. These assessments evaluate the benefits that ecosystems provide and identify how human activities impact those services. For example, an assessment of a wetland might quantify its role in filtering pollutants from water, reducing flood risks, and supporting biodiversity. By understanding these functions, policymakers can better appreciate the trade-offs involved in decisions such as converting wetlands for agriculture or development. Assessments provide a scientific basis for protecting valuable ecosystems and can guide land use planning, infrastructure projects, and conservation priorities.

Another valuable tool is natural capital accounting, which involves measuring and assigning economic value to natural assets, such as forests, rivers, and soils. Natural capital accounting helps policymakers recognize that ecosystems are not just passive backdrops for human activity but are critical economic assets that provide ongoing benefits. For instance, forests not only provide timber but also store carbon, support biodiversity, and regulate water cycles. By assigning monetary values to these services, natural capital accounting makes the economic importance of ecosystems visible and enables their inclusion in national accounts, alongside traditional measures of wealth like GDP. This shift encourages governments to invest in ecosystem protection and restoration as part of their economic planning.

Integrating ecosystem services into policy also has direct implications for sustainable development. For instance, urban planners who incorporate green infrastructure, such as parks, wetlands, and urban forests, into city designs can enhance the quality of life for residents while reducing the costs associated with flood damage, air pollution, and heat islands. Green spaces can improve public health by providing recreation areas and reducing the urban heat effect, while also supporting biodiversity and improving air quality. In coastal areas, preserving mangroves and coral reefs can protect communities from storm surges and support local fisheries, contributing to both economic resilience and ecological stability.

One major advantage of incorporating ecosystem services into policy and planning is its ability to help mitigate climate change and its impacts. Ecosystems such as forests, wetlands, and grasslands act as carbon sinks, absorbing carbon dioxide and helping regulate the global climate. By recognizing and protecting these ecosystem services, policymakers can create policies that align with climate goals while ensuring that communities benefit from the continued functioning of these natural systems. For example, policies that promote reforestation and protect existing forests can contribute to climate change mitigation efforts while also supporting biodiversity and local livelihoods.

In addition, ecosystem service integration can promote environmental justice by ensuring that marginalized communities, which often rely more heavily on ecosystem services, are not disproportionately affected by environmental degradation. For instance, indigenous communities may depend on forests and rivers for their livelihoods, culture, and food security. By integrating the value of these ecosystems into policy and planning, governments can protect these communities' rights and access to essential resources. This approach ensures that development projects are more equitable and that the benefits of ecosystem services are shared fairly across society.

Effective integration of ecosystem services into policy and planning requires cross-sector collaboration. Policymakers, environmental scientists, economists, and community stakeholders must work together to design policies that reflect the complexity of ecosystems and the interdependence of human and natural systems. For example, governments can collaborate with conservation organizations, the private sector, and local communities to develop strategies that protect ecosystems while promoting sustainable development. Engaging stakeholders in decision-making processes helps ensure that policies reflect the needs and values of those who depend on ecosystem services.

Incorporating the value of ecosystem services into policy and planning can significantly enhance conservation efforts and promote sustainable development. Tools such as ecosystem service assessments and natural capital accounting provide a framework for understanding the benefits ecosystems provide and ensuring that these benefits are considered in decision-making. By adopting this approach, policymakers can make more informed choices that support both environmental health and human prosperity, creating a more sustainable and resilient future for all.

11.3 Recognizing and Preserving Ecosystem Services

Recognizing and preserving ecosystem services is essential for maintaining clean air, water, pollination, and other vital functions that support human life and well-being. By protecting natural habitats, restoring degraded ecosystems, implementing sustainable practices, and promoting awareness, societies can ensure these services remain available for current and future generations. Effective management and conservation of ecosystem services are critical for achieving long-term environmental sustainability and improving quality of life.

When integrated and implemented effectively, the pathways to sustainability—such as promoting healthy lifestyles, addressing mental health impacts, recognizing and preserving ecosystem services, and advancing economic transformation—can guide humanity toward a more sustainable, equitable, and prosperous future. These strategies are interlinked and mutually reinforcing, creating a comprehensive framework for addressing the complex challenges of our time.

Promoting healthy lifestyles, including plant-based diets and active transportation, improves individual well-being and contributes to environmental sustainability. By encouraging habits that reduce environmental impact and enhance physical health, societies can create healthier populations and a healthier planet. Similarly, addressing mental health impacts related to environmental degradation and climate change helps build resilience, ensuring that people can cope with and adapt to the stresses of a changing world.

Recognizing and preserving ecosystem services is critical for maintaining the natural systems that support life. We safeguard the foundational elements that sustain human and ecological health by protecting clean air, water, and pollination services and implementing sustainable land and resource management practices. Integrating these services into policy and planning ensures that natural capital is valued and preserved for future generations.

Economic transformation, driven by green technologies and sustainable finance, creates new opportunities for growth while addressing environmental and social challenges. By fostering a green economy and promoting job creation in sustainable industries, societies can achieve economic prosperity that aligns with environmental stewardship and social equity.

These effective integration and implementation pathways offer a holistic approach to achieving sustainability. By collectively addressing health, mental well-being, ecosystem preservation, and economic transformation, humanity can advance toward a future where environmental health, social equity, and economic prosperity are harmoniously balanced (Harrison et al., 2014; Elmqvist et al., 2015; Carpenter et al., 2006; Martin-Ortega et al., 2015). This integrated approach enhances the quality of life for current and future generations and ensures that our planet's resources are managed responsibly and sustainably.

12

Conclusion

This book comprehensively explores the essence of Environmentalism-Pathways to Life for Humanity and the essential strategies for achieving sustainability in an increasingly complex world. It delves into the critical issues of environmental degradation, emphasizing the importance of sustainable development and environmental protection as foundational principles for a thriving future. The discussion extends to the *Sustainable Development Goals (SDGs)*, highlighting the necessity of balanced growth and resource efficiency and the pivotal role of renewable energy in climate change mitigation.

Key areas such as biodiversity conservation, social equity, and justice are addressed, underscoring the significance of accessible healthcare and the ethical imperative of inclusivity. The chapter also examines the transformative potential of technological innovation, green technology, and digital transformation in driving sustainable progress. The concept of a circular economy is presented as a vital approach to resource recycling and waste management, ensuring a sustainable product lifecycle.

Furthermore, the book underscores the importance of global cooperation, community engagement, and participatory governance in fostering shared knowledge and collaborative efforts. The exploration of economic transformation through green economy initiatives and sustainable finance illustrates how job creation and financial stability coexist with environmental stewardship. Finally, the chapter emphasizes the interconnectedness of health and well-being with ecosystem services, advocating for healthy lifestyles and mental health as integral components of a sustainable future.

This book covers these topics to inspire a new way of thinking and practicing sustainable living. It encourages readers to embrace holistic approaches that harmonize humanity's needs with the health of our planet.

13

The Future of Sustainability Principles and Practices

As the global community faces unprecedented environmental, social, and economic challenges, the future of sustainability principles and practices must evolve to meet these complex demands. Sustainability is no longer a niche concern; it has become a guiding framework for addressing climate change, social equity, biodiversity loss, resource depletion, and other global crises. The trajectory of sustainability research and practice will need to embrace a holistic and interdisciplinary approach, focusing not only on the mitigation of negative impacts but also on proactive, innovative solutions that can create a more resilient, just, and thriving future for all.

13.1 Focus Areas for Future Research

13.1.1 Climate resilience and adaptation: A critical path forward

As climate change continues to accelerate, the world is increasingly facing more frequent and severe environmental shocks—extreme weather events, rising sea levels, prolonged droughts, and shifting ecosystems. These changes are already having devastating impacts on ecosystems, human communities, and economies. To navigate this uncertain future, sustainability research must place a stronger emphasis on climate resilience and adaptation strategies. These strategies aim to build the capacity of both natural and human systems to withstand, recover from, and adapt to these growing threats.

13.1.2 Nature-based solutions for resilience

One of the most promising areas of research for climate resilience is the exploration of nature-based solutions (NbS). These are strategies that leverage the power of natural systems to address climate challenges. For example, restoring mangrove forests along coastlines can provide natural protection against storm surges, while also preserving biodiversity and enhancing carbon sequestration. Similarly, reforestation efforts in degraded landscapes not only improve ecosystems' ability to absorb carbon but also help maintain water cycles and soil health, which are critical for agricultural productivity. Research into NbS can help identify the most effective ways to harness the potential of nature in adapting to a rapidly changing climate, offering a sustainable and cost-effective alternative to traditional infrastructure-based solutions.

13.1.3 Regenerative agriculture as a path to climate adaptation

The agricultural sector is particularly vulnerable to climate change, with shifting weather patterns, rising temperatures, and extreme droughts threatening food security worldwide. Regenerative agriculture has emerged as a key approach for enhancing climate resilience in farming practices. This method goes beyond sustainability by actively restoring and improving soil health, increasing biodiversity, and enhancing water retention in the soil. Techniques like crop rotation, agroforestry, cover cropping, and reduced tillage all contribute to building ecosystems that are more resilient to climate shocks. Future research must delve deeper into the integration of these techniques across diverse agricultural landscapes, ensuring that farmers have access to the knowledge, tools, and resources they need to adopt regenerative practices. By promoting soil health and ecosystem restoration, regenerative agriculture offers a way to both mitigate climate change and build resilient food systems that can adapt to changing conditions.

13.1.4 Urban planning for climate-resilient cities

Urban areas, which house over half of the global population, are increasingly vulnerable to climate-related risks such as flooding, heatwaves, and resource shortages. Urban planning for climate resilience is an essential focus for both research and practice. Cities need to incorporate climate adaptation into their design and development, creating infrastructure that can cope with extreme weather conditions while also reducing their overall carbon footprint. Green infrastructure, such as urban forests, green roofs, and permeable pavements, can help mitigate the heat island effect, manage stormwater runoff, and improve air quality. Additionally, future research should explore how cities can be redesigned to be more energy-efficient, focusing on sustainable transportation systems, renewable energy integration, and efficient water management. Urban planners and policymakers must collaborate with climate scientists and engineers to build cities that are resilient in the face of escalating climate risks, ensuring the well-being of urban populations.

13.1.5 Innovations in renewable energy for resilience

Energy systems are central to both mitigating and adapting to climate change. The transition to renewable energy sources, such as solar, wind, and hydropower, is essential for reducing greenhouse gas emissions, but it also plays a critical role in enhancing resilience. Research into renewable energy innovations can uncover ways to make energy systems more flexible and adaptable in the face of climate-related disruptions. Decentralized energy systems, such as microgrids powered by renewables, are more resilient than centralized systems because they reduce reliance on a single power source and can continue functioning even when parts of the grid go down. Additionally, innovations in energy storage, such as advanced battery technologies, can ensure a stable energy supply even during periods of low wind or sunlight. As renewable energy becomes more integrated into the global energy landscape, research should focus on improving the efficiency, affordability, and accessibility of these technologies to ensure that communities and economies can thrive in a climate-resilient future.

13.1.6 Building community and economic resilience

Finally, climate resilience must also prioritize the social and economic dimensions of adaptation. Vulnerable communities—especially those in developing nations—are often the hardest hit by climate impacts due to their limited resources and infrastructure. Future research needs to focus on empowering these communities through education, resource distribution, and policy frameworks that promote adaptive capacity. Developing local solutions, such as early warning systems, community-based resource management, and climate-adapted housing, can significantly reduce vulnerability. Economically, there is a need to build more resilient industries and job markets that can withstand climate shocks. This includes diversifying income streams, developing climate-proof infrastructure, and encouraging green investments in renewable energy, sustainable agriculture, and climate-adapted urban development. Resilience is not just about bouncing back but about creating systems that are better equipped to handle future crises, ensuring long-term sustainability and equity.

13.1.7 An urgent research priority

The growing severity of climate change makes the need for climate resilience and adaptation strategies more urgent than ever. By prioritizing nature-based solutions, regenerative agriculture, climate-resilient urban planning, renewable energy innovations, and community resilience, sustainability research can help develop the tools and knowledge necessary for societies to adapt. The future of climate resilience will rely on a multidisciplinary approach that integrates environmental, social, and technological solutions, fostering a world where both people and nature can thrive in the face of an increasingly uncertain climate.

13.2 Circular Economy and Resource Efficiency: A Sustainable Shift for the Future

The traditional linear economy, based on the "take, make, dispose" model of production and consumption, is proving to be environmentally and economically unsustainable in the long term. This system leads to excessive resource extraction, significant waste generation, and environmental degradation. In contrast, the circular economy offers a transformative solution by promoting a closed-loop system in which resources are reused, recycled, and regenerated, thereby reducing the need for new raw materials and minimizing environmental impact.

13.2.1 The need for a circular economy

As global populations grow and consumption levels rise, the demand for finite natural resources is accelerating at an alarming rate. If left unchecked, this pattern will lead to resource scarcity, biodiversity loss, and worsening climate change. The circular economy represents a shift away from these unsustainable practices by focusing on resource efficiency, waste reduction, and the regeneration of natural systems. Unlike the linear model, the circular approach views waste not as an end-product but as a valuable resource that can be reintegrated into the economy. This requires innovative changes across the entire lifecycle of products, from design and manufacturing to consumption and disposal.

13.2.2 Innovation in material science and product design

One of the key drivers of a successful circular economy is innovation in material science and product design. To enable the efficient reuse and recycling of materials, products must be designed with longevity, repairability, and recyclability in mind. This includes using sustainable materials, reducing the number of components in a product, and making those components easy to separate and recycle. Advances in material science will also play a critical role in developing biodegradable and renewable alternatives to traditional materials such as plastics. Research into these areas will help businesses reduce their reliance on virgin raw materials while ensuring that products are designed to be part of a regenerative lifecycle.

For example, companies are already exploring the use of bio-based materials and recycled inputs for product manufacturing, which reduces the need for new resources. The development of materials that can be easily broken down and reused will further accelerate the shift toward circularity. Researchers and businesses must collaborate to discover new materials and processes that meet these sustainability goals, ensuring that products are designed for disassembly and reuse rather than disposal.

13.2.3 Sustainable waste management and recycling systems

Future research must also focus on advancing waste management and recycling systems to make the circular economy viable on a larger scale. Traditional waste management systems are often inadequate to handle the complexity of modern products, which contain a mix of materials that are difficult to separate. Developing more efficient and effective recycling processes is essential for turning waste into a resource that can be reintroduced into the production cycle.

Technological innovations such as chemical recycling, where materials are broken down into their basic chemical components and then rebuilt into new products, are promising solutions. Additionally, advances in waste sorting and collection infrastructure, powered by artificial intelligence and robotics, can improve the efficiency of recycling systems and increase the recovery rates of valuable materials. Research should focus on developing scalable solutions that can be implemented globally, particularly in regions where waste management infrastructure is underdeveloped.

13.2.4 Circular supply chain logistics

A transition to the circular economy also requires rethinking supply chain logistics. Traditional supply chains are designed for linear production, where raw materials are extracted, processed, and transformed into products that are used and eventually discarded. In a circular system, supply chains must be designed to facilitate the flow of reused and recycled materials back into the production process. This involves creating reverse logistics systems that collect used products, components, and materials, ensuring that they can be efficiently reintegrated into new production cycles.

Collaboration between industries, governments, and consumers is essential for creating an interconnected circular supply chain. Research should explore new business models that prioritize resource efficiency, such as product-as-a-service (where companies retain ownership of products and provide them as services to customers) or remanufacturing (where used products are restored to like-new condition). These models incentivize companies to design products that last longer and can be easily disassembled and repurposed.

13.2.5 Decoupling economic growth from environmental degradation

At the heart of the circular economy is the goal of decoupling economic growth from resource extraction and environmental degradation. Traditional economic growth has often been linked to increased resource use and environmental impact. However, the circular economy offers a pathway to sustainable growth by focusing on resource efficiency and the regeneration of natural systems. By reducing waste, minimizing resource use, and promoting the reuse of materials, the circular economy allows businesses to continue growing while minimizing their environmental footprint.

To achieve this, future research must focus on developing robust metrics to measure circularity and resource efficiency, ensuring that progress toward decoupling is transparent and measurable. Researchers should explore how circular business models can drive economic value while fostering environmental stewardship, creating opportunities for green jobs, innovation, and long-term sustainability.

The transition to a circular economy offers a blueprint for achieving sustainable economic growth that is no longer reliant on unsustainable resource extraction and waste generation. By advancing research in material science, product design, waste management, and supply chain logistics, we can create a system where resources are continually cycled back into the economy. This transformation will be critical in meeting the environmental challenges of the future, ensuring that economic prosperity can coexist with ecological balance. As businesses, governments, and researchers collaborate to implement circular practices, the future of sustainable development will be shaped by an economy that prioritizes resource efficiency, environmental stewardship, and long-term resilience.

13.3 Social Equity and Environmental Justice: A Cornerstone of Future Sustainability

Sustainability is not solely an environmental issue; it is deeply intertwined with social equity and justice. As the global community moves toward a more sustainable future, addressing the disproportionate impacts of environmental degradation on marginalized communities becomes a crucial element of sustainability. These communities, often already grappling with socio-economic challenges, tend to bear the brunt of climate change, pollution, and resource depletion. Therefore, achieving true sustainability requires embedding social equity and environmental justice at the heart of all sustainability initiatives.

13.3.1 The intersection of social inequalities and environmental issues

Environmental challenges are not experienced uniformly across society. Factors such as race, gender, and socioeconomic status profoundly influence how individuals and communities experience the impacts of environmental degradation. Marginalized groups—often poor, indigenous, or minority communities—frequently live in areas with higher exposure to pollution, lack of clean water, or limited access to natural resources. These groups are also more vulnerable to the impacts of climate change, such as extreme weather events, which exacerbate their already precarious living conditions.

Future sustainability research must prioritize intersectionality, a framework that examines how different aspects of identity—such as race, class, and gender—intersect to create unique experiences of disadvantage. By understanding these intersections, sustainability efforts can be tailored to address the specific needs and vulnerabilities of different communities. This approach ensures that marginalized voices are included in decision-making processes and that policies do not unintentionally deepen existing inequalities.

13.3.2 Environmental justice: Advocating for fair resource distribution

Environmental justice seeks to rectify the historical and ongoing inequalities that result in marginalized communities disproportionately bearing the negative effects of environmental degradation. A core aspect of environmental justice is the fair and equitable distribution of natural resources and environmental benefits. Many communities around the world, particularly in the Global South, suffer from resource scarcity while contributing the least to global environmental problems such as climate change.

Research should explore how policies can be designed to promote inclusive growth and equitable resource distribution. This includes not only ensuring that everyone has access to clean air, water, and other natural resources, but also addressing the socio-economic barriers that prevent certain groups from participating in the benefits of sustainable development.

Policy recommendations must incorporate principles of distributive justice, ensuring that marginalized communities receive a fair share of resources and opportunities to thrive in a sustainable future.

13.3.3 The role of policy and governance in advancing equity

Achieving social equity and environmental justice requires robust policy frameworks and governance structures that prioritize the needs of marginalized communities. Policies must be designed to mitigate the harmful environmental impacts on these communities, such as relocating waste facilities away from low-income areas, improving infrastructure resilience in vulnerable regions, and ensuring equitable access to green spaces and renewable energy sources.

Furthermore, governance systems must be inclusive, ensuring that marginalized groups are meaningfully involved in decision-making processes related to environmental and sustainability issues. This includes creating platforms for community engagement, where those most affected by environmental challenges can voice their concerns and participate in shaping the policies that affect their lives. By amplifying the voices of underrepresented communities, policymakers can create more inclusive and effective sustainability strategies.

A sustainable future must be one in which economic growth is inclusive and benefits all members of society. Inclusive growth focuses on ensuring that the economic opportunities generated by sustainable development are accessible to marginalized groups, including women, indigenous peoples, and low-income communities. Research in this area should explore strategies for creating green jobs and training programs that empower these communities to participate in the growing green economy.

Additionally, equitable access to education and resources will be critical for enabling marginalized groups to thrive in a sustainable future. Ensuring that all communities have access to sustainability education, training, and technology will help close the gap between those who benefit from sustainability initiatives and those who are left behind. By fostering inclusivity in both environmental and economic realms, we can create a more just and sustainable world.

The future of sustainability must prioritize social equity and environmental justice to ensure that no one is left behind in the transition to a sustainable world. By addressing the deep-rooted inequalities that disproportionately affect marginalized communities, sustainability research and policy can create a more inclusive and just society. This requires a commitment to intersectionality, fair resource distribution, and inclusive governance structures. As we move forward, sustainability must be redefined not only in terms of environmental protection but also as a pathway to social justice and equity for all.

13.4 Technological Innovation for Sustainability: Harnessing the Power of New Technologies

The rapid evolution of technology presents unprecedented opportunities for driving sustainability across multiple sectors. Innovations in artificial intelligence (AI), biotechnology, data analytics, and other emerging technologies offer new tools to enhance resource efficiency, monitor environmental changes, and develop more sustainable industrial practices. As humanity faces escalating environmental challenges, integrating these cutting-edge technologies can significantly bolster efforts toward a sustainable future.

13.4.1 Optimizing resource use through advanced technologies

One of the most promising aspects of technological innovation is its potential to optimize resource use, reducing waste and promoting more sustainable consumption and production patterns. Artificial intelligence and data analytics enable more accurate modeling of resource flows, allowing businesses and governments to fine-tune their operations to minimize environmental impacts. For example, AI-driven smart grids can optimize energy distribution, reducing waste and improving the efficiency of renewable energy sources like solar and wind power. Similarly, precision agriculture—using drones, sensors, and AI—can optimize water and fertilizer use, leading to more sustainable food production while conserving natural resources.

In industrial processes, machine learning algorithms can identify inefficiencies, propose improvements, and streamline operations to reduce carbon footprints. These innovations hold immense potential to drive sustainability across diverse sectors, from manufacturing to transportation and energy production. However, future research is needed to explore how these technologies can be scaled up and integrated into existing systems to achieve widespread sustainability benefits.

13.4.2 Monitoring ecosystems and predicting environmental risks

Emerging technologies also provide powerful tools for monitoring ecosystems and predicting environmental risks, enabling more proactive and effective responses to ecological challenges. Remote sensing technologies, coupled with AI, allow for real-time monitoring of deforestation, land-use changes, and pollution levels, providing critical data for conservation efforts. These technologies can help track biodiversity, monitor ecosystem health, and assess the impacts of human activities on natural habitats, offering invaluable insights for policymakers and conservationists.

Additionally, predictive analytics powered by AI can model climate risks and natural disasters, such as floods, hurricanes, and droughts, allowing for early warning systems and more resilient urban planning. By providing accurate forecasts of environmental threats, communities and governments can take preemptive measures to mitigate damage and protect vulnerable populations. As climate change accelerates the frequency of extreme weather events, these technologies will play a vital role in building climate resilience and safeguarding ecosystems.

13.4.3 Driving sustainable industrial practices

Technological innovations are also essential for driving sustainable industrial practices. Biotechnology, for instance, offers breakthroughs in creating sustainable materials, such as biodegradable plastics and biofuels, which can replace traditional, environmentally harmful alternatives. 3D printing technologies can revolutionize manufacturing by reducing material waste and enabling on-demand production, minimizing the need for excessive transportation and inventory.

Moreover, blockchain technology can be utilized to increase transparency in supply chains, ensuring that products are sourced ethically and sustainably. This can help reduce illegal practices such as deforestation and overfishing, as blockchain enables accurate tracking of goods from their origin to the consumer. Future research should focus on further developing and integrating these technologies into sustainable industrial frameworks, fostering the creation of green supply chains and circular production models.

13.4.4 Ethical considerations in technological deployment

While the potential benefits of technological innovation for sustainability are vast, the ethical considerations surrounding their deployment must not be overlooked. Questions about data privacy, resource allocation, and the unintended environmental consequences of new technologies are crucial topics for future research. For example, AI and big data systems require significant energy to process information, which can generate a large carbon footprint if not managed properly. The production of advanced technologies, such as batteries for electric vehicles, also relies on finite resources like lithium, raising concerns about resource extraction and environmental degradation.

It is essential that future studies not only focus on the technical feasibility of these innovations but also critically assess their social and environmental impacts. Researchers should explore frameworks for ethical technology deployment that prioritize equity, inclusivity, and long-term sustainability. This includes ensuring that new technologies are accessible to all communities, particularly those most vulnerable to environmental change, and that their implementation contributes to the greater social and ecological good.

The future of sustainability is closely linked to the responsible application of technological innovations. As AI, biotechnology, and other advanced technologies continue to evolve, they offer immense potential to optimize resource use, enhance environmental monitoring, and drive sustainable industrial practices. However, careful attention must be paid to the ethical dimensions of these technologies, ensuring that they are used in ways that promote social equity and environmental justice. Future research must continue to explore how these technologies can be harnessed to create a more sustainable world while ensuring that their deployment is guided by ethical considerations for the benefit of all.

13.5 Behavioral Change and Sustainability Culture: Shaping the Future through Human Action

While technological advancements are crucial to achieving sustainability, technical solutions alone are insufficient. Human behavior and cultural values play a critical role in driving the transition toward a sustainable future. Without shifts in how individuals, businesses, and governments approach environmental challenges, even the most advanced technologies may fall short of their potential. The future of sustainability research must increasingly focus on understanding behavioral change and fostering a culture of sustainability that encourages long-term, systemic shifts.

13.5.1 The role of behavioral science in sustainability

Behavioral science offers vital insights into why people make certain choices and how they can be influenced to make more sustainable ones. Research in this field explores the psychological, social, and economic factors that motivate or discourage environmentally friendly behavior. For example, understanding how people perceive climate risks, value environmental goods, or prioritize short-term gains over long-term sustainability can inform strategies to promote more responsible consumption patterns.

Future research should delve deeper into the psychological barriers to sustainability, such as cognitive dissonance, status quo bias, and social norms that encourage unsustainable behavior. By identifying these barriers, sustainability advocates can design targeted interventions that encourage positive changes. For instance, nudging—a concept from behavioral economics—can be applied to prompt individuals toward more sustainable choices, like conserving energy or reducing waste, by making these options easier or more attractive.

13.5.2 Education, communication, and policy as drivers of change

To foster widespread behavioral change, education and communication must be central components of sustainability initiatives. Environmental education can shape the attitudes and values of younger generations, equipping them with the knowledge and skills needed to address environmental challenges. Likewise, effective communication campaigns can raise awareness about sustainability issues and inspire action across different sectors of society.

Future research should examine the most effective methods of communication for reaching diverse audiences, including marginalized communities, policymakers, and corporate leaders. Studies might focus on how messaging can be tailored to specific cultural contexts or how to best leverage digital media and social platforms to amplify sustainability efforts.

Beyond education and communication, policy incentives also play a critical role in promoting behavioral change. Research is needed to explore which types of policies are most effective at encouraging sustainable practices at both the individual and institutional levels. For instance, carbon pricing, subsidies for renewable energy, or tax incentives for eco-friendly businesses can create the financial motivations necessary for widespread adoption of sustainability measures.

13.5.3 Building a sustainability culture

A truly sustainable future requires more than just individual actions; it demands the development of a sustainability culture that permeates all aspects of society. This means embedding environmental values into everyday practices, business operations, and government policies. Cultural change is often slow, but when sustainability becomes a shared social value, it has the power to reshape norms, priorities, and behaviors.

Research into cultural shifts should explore how societies can transition from consumer-driven, growth-oriented models to ones emphasizing environmental stewardship, social equity, and community well-being. Case studies from different parts of the world, where sustainability is already a core cultural value, can provide insights into how to replicate such shifts in other regions.

13.5.4 The intersection of individual and collective action

The relationship between individual behavior and collective action is another critical area for future research. While personal choices—such as reducing waste, conserving energy, or adopting plant-based diets—are important, they must be complemented by broader systemic changes. Research should explore how individual actions can scale up to collective movements and influence institutional and policy changes.

For instance, social movements and grassroots organizations play an important role in pushing for larger-scale sustainability initiatives, from corporate accountability to government policy reforms. Understanding how individuals become part of these movements and what motivates collective action can help drive larger societal shifts toward sustainability.

The path to sustainability must be paved with both technological advancements and significant behavioral shifts. Future research should focus on understanding the complex factors that influence human behavior and designing effective interventions to foster a culture of sustainability. By integrating education, communication, and policy incentives with an understanding of behavioral science, the sustainability movement can engage people, businesses, and governments in meaningful, long-lasting change. Only through this combined approach will we be able to address the environmental challenges of the 21st century and beyond.

13.6 Integration of Sustainability into Policy and Governance: A Pathway to Holistic Change

The integration of sustainability into policy and governance is critical for addressing the complex and interconnected challenges of our time. As the impacts of climate change, biodiversity loss, and social inequalities become increasingly evident, governments and policymakers must embed sustainability into their regulatory frameworks and national development strategies. This requires a fundamental shift in how policies are conceived, designed, and implemented, ensuring they align with global sustainability goals, such as the United Nations' Sustainable Development Goals (SDGs).

13.6.1 Establishing clear and enforceable standards

To effectively incorporate sustainability, policymakers must establish clear, enforceable standards that serve as benchmarks for progress. This includes developing comprehensive regulations that address environmental protection, social equity, and economic viability. By setting measurable targets, governments can hold themselves accountable and facilitate transparency in the implementation of sustainability initiatives. Moreover, such standards can drive innovation within industries, encouraging businesses to adopt sustainable practices and technologies.

For instance, establishing stricter emissions standards for industries can incentivize companies to invest in cleaner technologies and reduce their carbon footprints. Similarly, regulations that promote sustainable agriculture practices can help conserve biodiversity and improve food security. By creating a regulatory environment that prioritizes sustainability, governments can lead by example and signal to all stakeholders the importance of these values in decision-making.

13.6.2 Multi-stakeholder collaboration: A key to success

The integration of sustainability into policy and governance cannot be achieved in isolation; it necessitates multi-stakeholder collaboration. Engaging a diverse range of actors—including governments, the private sector, civil society, and local communities—ensures that policies are inclusive, equitable, and reflective of various interests and needs. Collaborative approaches facilitate the sharing of knowledge, resources, and best practices, which can enhance the effectiveness of sustainability initiatives.

For example, partnerships between governments and businesses can drive innovations in sustainable technologies while ensuring that regulatory frameworks do not stifle economic growth. Civil society organizations can play a crucial role in advocating for marginalized communities and ensuring that their voices are heard in policy discussions. By fostering collaboration across sectors, governments can create a more holistic approach to sustainability that considers the complexities of social, environmental, and economic systems.

13.6.3 Aligning with global sustainability goals

Integrating sustainability into policy also means aligning national strategies with global sustainability goals, particularly the SDGs. Countries must assess their development plans through the lens of these goals, identifying opportunities to advance sustainability while addressing local needs. This alignment can provide a roadmap for countries to navigate the challenges they face while contributing to a more sustainable global community.

Moreover, by integrating sustainability into national policies, countries can enhance their credibility and competitiveness on the global stage. Nations that take proactive steps toward sustainability are better positioned to attract investment, access international markets, and participate in global initiatives aimed at addressing pressing environmental and social issues.

13.6.4 Challenges and opportunities Ahead

Despite the clear benefits of integrating sustainability into policy and governance, challenges remain. Political resistance, bureaucratic inertia, and competing interests can impede progress. However, these challenges present an opportunity for advocates of sustainability to engage in constructive dialogue with policymakers, emphasizing the long-term benefits of sustainable practices for society, the economy, and the environment.

Furthermore, the ongoing climate crisis and social inequalities provide a powerful impetus for change. As public awareness grows and communities demand action, governments have a unique opportunity to respond by integrating sustainability into their policies and demonstrating leadership in global efforts to tackle these pressing issues.

The integration of sustainability into policy and governance is not merely a trend; it is an essential pathway toward a sustainable future. By establishing clear standards, fostering multi-stakeholder collaboration, and aligning with global sustainability goals, governments can create a regulatory framework that promotes sustainability and resilience. As we confront the myriad challenges of our time, the role of policymakers will be crucial in shaping a world where environmental protection, social equity, and economic viability are intertwined, ensuring a better future for generations to come.

13.7 Corporate Responsibility and Sustainable Business Models: Shaping the Future of Business

As we move into an era increasingly defined by environmental challenges and social inequities, the integration of sustainability into corporate strategy will become essential for long-term success. Corporate responsibility is evolving beyond traditional corporate social responsibility (CSR) initiatives, requiring businesses to adopt sustainable business models that prioritize ecological integrity and social well-being over mere profit. This paradigm shift recognizes that businesses are not isolated entities; rather, they are integral components of broader societal and environmental systems.

13.7.1 Embedding sustainability in corporate strategy

The future of corporate sustainability lies in embedding sustainable practices into the very fabric of business operations and strategies. This means that sustainability will no longer be a peripheral concern but a central tenet of how companies operate and create value. Businesses must assess their environmental impacts, identify areas for improvement, and implement strategies that contribute to a sustainable future. This includes investing in green technologies, such as renewable energy sources and sustainable materials, to minimize their ecological footprints.

Moreover, companies must actively reduce their carbon emissions by adopting energy-efficient practices and transitioning to renewable energy. By setting ambitious carbon reduction targets and transparently reporting on progress, businesses can comply with regulations and build trust and loyalty among stakeholders.

13.7.2 Ensuring supply chain transparency

Supply chain transparency will also be a critical component of sustainable business models. Companies must ensure that their supply chains are not only efficient but also ethical and sustainable. This includes scrutinizing the sourcing of raw materials, ensuring fair labor practices, and minimizing environmental impacts throughout the supply chain. By adopting practices such as traceability and responsible sourcing, businesses can foster consumer trust and mitigate risks associated with unethical practices.

Additionally, companies should collaborate with suppliers to implement sustainability initiatives, share best practices, and develop innovative solutions that drive sustainability across the supply chain. This collaborative approach can enhance resilience and create shared value for all stakeholders involved.

13.7.3 Responding to consumer demand for ethical products

The landscape of consumer preferences is shifting dramatically, with growing demand for ethical products and services. Consumers are increasingly aware of the environmental and social implications of their purchases, leading them to seek out companies that demonstrate a commitment to sustainability. This trend presents both a challenge and an opportunity for businesses. Companies that fail to adapt risk losing market share, while those that embrace sustainability can differentiate themselves and build brand loyalty.

To meet this demand, businesses must not only offer sustainable products but also communicate their sustainability efforts effectively. Transparency in marketing and reporting can enhance credibility and foster a sense of connection with consumers. By sharing stories about sustainable sourcing, environmental initiatives, and community engagement, companies can engage consumers and encourage them to become advocates for sustainable practices.

13.7.4 Shaping the future of corporate sustainability

As sustainability becomes more deeply embedded in corporate strategy, businesses will face new challenges and opportunities. The adoption of sustainable business models will require a cultural shift within organizations, emphasizing innovation, collaboration, and accountability. Companies will need to invest in research and development to explore new technologies and practices that enhance sustainability.

Furthermore, regulatory pressures and stakeholder expectations will continue to rise, pushing companies to set ambitious sustainability goals and report on their progress transparently. In this context, corporate leaders must recognize that sustainability is not just a compliance issue but a fundamental driver of long-term business success.

The future of sustainability practices in the corporate world will hinge on the ability of businesses to redefine their roles and responsibilities. By embracing sustainable business models that prioritize long-term ecological and social well-being, companies can contribute to a more sustainable future while enhancing their resilience and competitiveness. As stakeholders increasingly demand ethical and responsible practices, businesses that lead in sustainability will not only thrive but also play a vital role in shaping a more equitable and sustainable world for all.

13.8 Community-led Sustainability Initiatives: Empowering Local Action for a Sustainable Future

In an era marked by escalating environmental challenges, community-led sustainability initiatives have emerged as vital forces for change, promoting sustainable practices at the grassroots level. These initiatives prioritize the active involvement of local communities, recognizing that sustainable solutions must be tailored to the specific needs and contexts of each community.

By empowering communities to take charge of their natural resources and environment, we can foster resilience and adaptability in the face of climate change and resource depletion.

13.8.1 Empowering communities through cooperative farming

One prominent example of community-led sustainability is cooperative farming, where local farmers collaborate to manage resources, share knowledge, and enhance productivity. By pooling resources and expertise, communities can adopt sustainable agricultural practices that not only increase food security but also minimize environmental impacts. Cooperative farming allows for the implementation of agroecological practices, such as crop rotation and organic farming, which promote biodiversity, improve soil health, and reduce reliance on chemical inputs.

Moreover, cooperative models often prioritize local consumption, fostering connections between farmers and consumers, and strengthening local economies. This approach not only enhances resilience to market fluctuations but also encourages a sense of community ownership and responsibility toward sustainable practices.

13.8.2 Local renewable energy projects

Local renewable energy projects are another critical component of community-led sustainability initiatives. As communities seek to reduce their reliance on fossil fuels and lower their carbon footprints, grassroots efforts to develop renewable energy sources such as solar, wind, and biomass are gaining momentum. Community-owned energy projects enable residents to harness local resources and generate clean energy, thereby reducing energy costs and promoting energy independence.

These initiatives often incorporate educational components, raising awareness about the benefits of renewable energy and encouraging sustainable energy consumption. By engaging community members in designing and implementing energy projects, these initiatives foster a sense of ownership and collective responsibility for environmental stewardship.

13.8.3 Participatory governance models

Participatory governance models are essential for ensuring that communities have a voice in decision-making processes regarding the management of local natural resources. These models promote transparency, accountability, and inclusivity by involving community members in governance structures.

Such approaches can include participatory budgeting, community forums, and stakeholder councils, allowing residents to influence policies and practices directly impacting their environment and livelihoods.

Empowering communities to participate in governance enhances local ownership of sustainability initiatives and builds trust between community members and decision-makers. This collaborative approach fosters a sense of shared responsibility, encouraging community members to actively implement sustainable practices and advocate for their needs.

13.8.4 Adapting to environmental changes

Community-led sustainability initiatives provide a framework for resilience and adaptation in the face of environmental changes. By leveraging local knowledge and expertise, communities can develop context-specific strategies that address their unique challenges. These may include disaster preparedness plans, water conservation initiatives, and habitat restoration projects that enhance ecosystem resilience.

Communities that are empowered to manage their resources sustainably are better equipped to respond to environmental changes and recover from shocks. Through collaboration and collective action, these initiatives foster social cohesion and build a sense of community resilience, ensuring that communities can thrive in an increasingly uncertain future.

As we look to the future, the importance of community-led sustainability initiatives cannot be overstated. By prioritizing local action, these initiatives not only address pressing environmental challenges but also empower communities to shape their destinies. By fostering cooperative farming, developing local renewable energy projects, and promoting participatory governance, communities can become active agents of change in the pursuit of sustainability.

In doing so, they lay the groundwork for a more equitable and sustainable future, where the voices and needs of local populations are at the forefront of decision-making. Through these grassroots efforts, we can create a more resilient world, capable of adapting to the complex challenges posed by climate change and environmental degradation.

13.9 Global Collaboration and Knowledge Sharing: A Unified Approach to Sustainability Challenges

In an increasingly interconnected world, global collaboration and knowledge sharing are essential components for effectively addressing sustainability challenges that transcend national borders. Issues such as climate change, biodiversity loss, and environmental degradation do not respect geopolitical boundaries; they demand coordinated international responses that leverage the collective expertise and resources of diverse stakeholders. The future of sustainability practices must prioritize collaboration among governments, researchers, non-governmental organizations (NGOs), and local communities to create comprehensive solutions that benefit the planet and its inhabitants.

13.9.1 The need for international cooperation

Climate change and biodiversity loss pose significant threats to ecosystems and human well-being worldwide. These challenges require a united front, as individual nations alone cannot solve problems that are inherently global. International agreements, such as the Paris Agreement, exemplify the importance of nations coming together to set collective goals for emissions reductions and conservation efforts. However, to translate these agreements into meaningful action, there must be a concerted effort to foster collaboration across disciplines, sectors, and geographical regions.

13.9.2 Platforms for knowledge sharing

To facilitate effective global collaboration, robust platforms for knowledge sharing are imperative. International conferences and research networks serve as crucial venues for disseminating research findings, best practices, and innovative solutions. Such gatherings enable stakeholders to exchange ideas, discuss challenges, and forge partnerships that enhance the collective understanding of sustainability issues. By bringing together experts from various fields—ranging from ecology and economics to social sciences—these platforms encourage interdisciplinary approaches that are vital for tackling complex sustainability challenges.

Moreover, open-access databases and digital repositories can significantly enhance the accessibility of information and research findings. By making data freely available, stakeholders can leverage existing knowledge, avoid duplicative efforts, and build upon the successes of others. This democratization of knowledge fosters an inclusive environment where all voices can contribute to the sustainability discourse, particularly those from marginalized communities who often possess invaluable local knowledge and experiences.

13.9.3 Innovative solutions through collaboration

Global collaboration also opens avenues for developing innovative solutions that may not be possible in isolation. For example, collaborative research initiatives can harness diverse perspectives and expertise to address specific sustainability challenges more effectively. By pooling resources and knowledge, researchers can undertake large-scale studies, pilot projects, and field experiments that yield insights into the most effective strategies for resilience and adaptation.

Furthermore, public-private partnerships can play a crucial role in fostering innovation. By engaging businesses, governments, and civil society organizations, these collaborations can leverage funding, technological advancements, and grassroots knowledge to implement sustainable practices at scale. As companies increasingly recognize the importance of sustainability, their involvement in global initiatives can accelerate the transition to more sustainable business models, thereby amplifying the impact of collaborative efforts.

13.9.4 Building a culture of collaboration

To embed global collaboration into the fabric of sustainability practices, it is essential to foster a culture of collaboration and openness. This requires not only establishing formal networks and platforms but also promoting values of inclusivity, respect, and shared responsibility among all stakeholders. Education and capacity-building initiatives can empower individuals and organizations to engage in collaborative efforts effectively, equipping them with the skills and knowledge needed to navigate complex global challenges.

As we look toward the future, the emphasis on global collaboration and knowledge sharing will be critical in addressing the multifaceted sustainability challenges facing humanity. By fostering international cooperation, creating robust platforms for knowledge exchange, and embracing innovative solutions, we can harness the collective power of diverse stakeholders to drive meaningful progress toward a sustainable future.

In doing so, we must recognize that sustainability is not just an environmental issue but a shared human endeavor that requires collaboration across borders, sectors, and disciplines. Through concerted global efforts, we can pave the way for a more resilient and sustainable world for generations to come.

13.10 Navigating the Future of Sustainability

The future of sustainability principles and practices requires a paradigm shift characterized by creativity, adaptability, and inclusivity. As we confront the complex challenges posed by climate change, resource depletion, and social inequity, it is evident that traditional approaches are insufficient. Researchers, policymakers, businesses, and communities must collaboratively devise strategies that not only mitigate the symptoms of unsustainability but also address the systemic issues at their core. This transformative approach necessitates a deep understanding of the interconnectedness of environmental, social, and economic systems and a commitment to reimagining how we live, work, and interact with the world around us.

To navigate this future effectively, the sustainability domain will increasingly depend on interdisciplinary collaboration. Bringing together diverse fields of expertise—such as environmental science, economics, sociology, and technology—enables a holistic understanding of sustainability challenges. By fostering collaboration among researchers, practitioners, and stakeholders from different sectors, we can develop comprehensive solutions that are informed by a wide range of perspectives and experiences. This collaborative spirit will not only enhance the effectiveness of sustainability initiatives but also encourage innovation and creative problem-solving, leading to more impactful outcomes.

Technological innovation will play a crucial role in shaping the future of sustainability practices. Emerging technologies, such as artificial intelligence, big data analytics, and renewable energy solutions, have the potential to revolutionize how we approach sustainability. These technologies can optimize resource use, improve efficiency, and provide new ways to monitor and manage environmental impacts. However, the deployment of these innovations must be guided by ethical considerations and a commitment to social equity, ensuring that technological advancements benefit all segments of society, particularly those who are often marginalized or disproportionately affected by environmental challenges.

A strong commitment to social equity will be essential in creating a sustainable future where all individuals and communities can thrive. Addressing deep-rooted inequalities and ensuring that marginalized voices are heard will lead to more inclusive and effective sustainability solutions. Research must delve into the intersectionality of environmental issues and social justice, examining how factors such as race, gender, and socioeconomic status influence access to resources and opportunities. By prioritizing equity in sustainability practices, we can build a more just and resilient society that uplifts all individuals, regardless of their background.

As we shift our focus toward resilience, circular economies, and innovation, we lay the groundwork for a sustainable future that is both just and prosperous. Resilience emphasizes the importance of preparing for and adapting to environmental changes, ensuring that communities can withstand shocks and thrive in the face of adversity. Circular economies promote sustainable production and consumption patterns that minimize waste and maximize resource efficiency, fundamentally altering how we view economic growth. By fostering a culture of innovation, we can continually evolve our approaches to sustainability, harnessing creativity and new ideas to tackle emerging challenges.

13.11 Proposing New Ways to Manage Humanitarian and Planetary Issues

Given the immense challenges outlined above, it is clear that achieving the SDGs by 2030, under the current system, is highly unlikely. Therefore, new approaches and paradigms must be considered to manage both humanitarian and planetary environmental issues.

13.11.1 Localized solutions with global support

Instead of relying on global one-size-fits-all policies, development strategies should focus on localized solutions tailored to the specific needs and contexts of individual regions and communities. This means empowering local communities with the resources, knowledge, and technology they need to pursue their own sustainable development pathways. However, global cooperation is essential to ensure that these local efforts are supported through funding, technological transfers, and capacity-building initiatives. A bottom-up approach, where communities lead the charge in setting priorities and goals, could help achieve more meaningful and contextually relevant progress.

13.11.2 Reimagining economic systems

The current economic model, which prioritizes growth over sustainability, is unsuited to the long-term goals of sustainable development. A shift toward circular economies that focus on minimizing waste, maximizing resource efficiency, and promoting regenerative practices is necessary. Additionally, ecological economics, which values natural resources and ecosystem services, should be mainstreamed into global economic frameworks. This would require redefining indicators of progress, moving away from GDP as the primary measure of success, and instead focusing on metrics such as well-being, resource regeneration, and environmental health.

13.11.3 Global climate and environmental governance

Climate change requires a global governance system capable of enforcing sustainability measures across national borders. A new international framework that goes beyond voluntary agreements like the Paris Agreement is necessary. This framework could include enforceable environmental treaties, sanctions for countries that fail to meet their climate commitments, and incentives for those that adopt aggressive sustainability policies. Climate resilience should also be integrated into all national and regional policies to ensure that long-term strategies for adaptation are in place.

13.11.4 Increased investment in green technologies

Scaling up investment in green technologies is essential to meet the SDGs. This includes increasing funding for research and development in renewable energy, sustainable agriculture, and climate mitigation technologies. Moreover, ensuring that these technologies are accessible to all, particularly in the Global South, will require innovative financing mechanisms such as green bonds, carbon credits, and public-private partnerships. Governments should also incentivize businesses to invest in sustainability through subsidies and tax breaks for green technologies and sustainable practices.

13.11.5 Redefining global aid and financing structures

Global aid and financing structures need to be redefined to focus on the long-term sustainability of countries, rather than short-term developmental aid. A new system of sustainable development financing, which prioritizes climate resilience, environmental protection, and equitable growth, should be established. This may include debt forgiveness for developing nations in exchange for investments in sustainability, as well as increased global taxes on carbon emissions and large corporations to fund SDG efforts. These reforms could ensure that financial resources are aligned with the global need for sustainable development.

13.11.6 Fostering behavioral and cultural change

Achieving sustainable development will also require profound behavioral and cultural shifts. Public awareness campaigns, education, and the promotion of sustainability as a core cultural value will be necessary to drive changes in consumption patterns, energy use, and lifestyle choices. Governments, civil society, and businesses must work together to create incentives for sustainable living while discouraging unsustainable behaviors through regulations and taxes on environmentally harmful products and practices.

The challenges to achieving the Sustainable Development Goals are vast and multifaceted, but they are not insurmountable. By rethinking economic systems, fostering localized solutions, enhancing global governance, and increasing investment in green technologies, we can move closer to a future where both humanitarian and planetary environmental issues are addressed. While the 2030 timeline may seem unrealistic under the current trajectory, the pursuit of these goals remains critical for building a just, resilient, and sustainable world for future generations.

13.12 Sustainable Development Goals: Why Achieving Them Seems Impossible and How We Can Turn the Impossible into Possible

The Sustainable Development Goals (SDGs), established by the United Nations in 2015, represent a global commitment to address critical social, economic, and environmental challenges by 2030. These 17 goals, ranging from ending poverty and hunger to achieving gender equality and climate action, are ambitious in scope and essential for the future of humanity and the planet. However, despite the global effort and enthusiasm surrounding the SDGs, many argue that achieving them within the given timeframe seems impossible. This sentiment arises from a number of significant practical, political, economic, and systemic barriers.

13.12.1 Why achieving the SDGs seems impossible

13.12.1.1 Financial shortfalls and inequality

One of the most prominent challenges to achieving the SDGs is the lack of sufficient financial resources. Implementing the SDGs on a global scale requires trillions of dollars annually, but current financial commitments fall short of these needs. Developing nations, in particular, struggle to find the necessary funding to improve infrastructure, healthcare, education, and clean energy. Additionally, global inequality further compounds this challenge, as wealthier countries and regions have the resources to make progress, while poorer nations fall behind, widening the development gap.

13.12.1.2 Political will and governance challenges

Achieving the SDGs requires strong political will and governance across all levels—from local communities to national governments and international organizations. Unfortunately, many countries face political instability, corruption, or competing priorities that detract from efforts to meet these goals.

Political leaders may prioritize short-term gains over long-term sustainability, resulting in policies that undermine progress on the SDGs. Furthermore, differences in political ideologies and governance structures across nations make it difficult to implement coherent, unified strategies globally.

13.12.1.3 Climate change and environmental pressures

Climate change is one of the most significant barriers to achieving the SDGs. Its widespread effects—ranging from more frequent natural disasters to shifts in agricultural productivity—have already derailed progress in key areas like poverty reduction, food security, and biodiversity preservation. Climate-related risks disproportionately affect the poorest and most vulnerable communities, making it even more challenging to reach goals related to social equity and economic growth. Without aggressive global climate action, the SDGs will remain unattainable.

13.12.1.4 Economic systems and unsustainable growth

The current global economic model, which prioritizes short-term profit and continuous growth, directly conflicts with many SDGs, especially those related to environmental sustainability. Economic systems are heavily reliant on resource extraction, overconsumption, and fossil fuels, contributing to environmental degradation and exacerbating inequality. As long as unsustainable economic practices dominate global markets, achieving the SDGs, particularly those focused on responsible consumption and production, will remain a distant goal.

13.12.1.5 Social and cultural barriers

Achieving the SDGs also requires fundamental shifts in social and cultural behaviors. Many of the goals, such as those promoting gender equality, education, and sustainable living, involve changes in deeply ingrained societal norms and values. These changes are difficult to implement on a large scale, particularly in regions where traditional customs and practices conflict with SDG objectives. Overcoming cultural resistance requires significant investments in education, awareness, and policy enforcement.

13.12.2 How to turn the impossible into possible

While the obstacles to achieving the SDGs are daunting, they are not insurmountable. There are several strategies and innovations that can make progress possible, even if the 2030 timeline seems ambitious.

13.12.2.1 Innovative financing for sustainable development

One of the key ways to address the financial gap is through innovative financing mechanisms. This includes leveraging public-private partnerships, green bonds, and impact investing, which direct capital toward projects that contribute to the SDGs. Additionally, debt-for-sustainability swaps could help developing nations reduce their financial burdens while funding sustainable projects. Global financial institutions must also be restructured to prioritize long-term sustainability over short-term profit, ensuring that investments are aligned with the SDGs.

13.12.2.2 Strengthening global governance and accountability

Global governance systems need to be strengthened to create enforceable international standards for SDG progress. A new framework could include binding treaties and agreements that hold countries accountable for their commitments to sustainability, human rights, and climate action. To ensure political will, international bodies like the UN, World Bank, and IMF could tie financial aid and development assistance to demonstrated progress toward SDGs. Transparent reporting systems and regular assessments of national and regional progress would help identify gaps and hold governments accountable.

13.12.2.3 Localized solutions and community empowerment

The SDGs should not be pursued through top-down, one-size-fits-all approaches. Instead, they must be localized to reflect the specific needs and challenges of individual countries and communities. Local communities should be empowered to take ownership of SDG initiatives through participatory governance, ensuring that development strategies are contextually appropriate and grounded in local knowledge. Grassroots organizations, indigenous communities, and local stakeholders should play a central role in co-designing and implementing sustainable solutions that work for their regions.

13.12.2.4 Technological innovation and data-driven solutions

Technology holds great promise for making the SDGs achievable. Artificial intelligence, big data, blockchain, and biotechnology can be harnessed to solve some of the most pressing challenges, such as climate adaptation, renewable energy, healthcare, and education. For example, smart agriculture can improve food security by optimizing crop yields in a changing climate, while clean energy technologies can provide sustainable power to underserved communities. Additionally, real-time data and predictive analytics can help governments and organizations make informed decisions and respond quickly to emerging challenges.

13.12.2.5 Circular economy and sustainable business practices

Transforming the global economy into a circular economy is essential for achieving the SDGs. This model emphasizes the reuse, recycling, and regeneration of resources, rather than the linear model of consumption and waste. Businesses must adopt sustainable practices that prioritize long-term ecological and social well-being over short-term profits. This shift requires corporate accountability for environmental impacts, transparency in supply chains, and investment in green technologies. Consumer demand for ethical products and services will also push companies toward more sustainable practices, making the business case for SDG-aligned strategies stronger.

13.12.2.6 Climate action and resilience building

Climate action is critical for achieving many of the SDGs, particularly those related to poverty, hunger, water, and ecosystems. Immediate and aggressive action is needed to mitigate the effects of climate change, such as transitioning to renewable energy, reducing carbon emissions, and protecting natural ecosystems. Building climate resilience through nature-based solutions, climate-smart agriculture, and resilient infrastructure is essential for ensuring that communities can adapt to the impacts of climate change and thrive in the long term.

13.12.2.7 Promoting behavioral change and sustainable lifestyles

Achieving the SDGs requires a global cultural shift toward sustainability. Governments, civil society, and businesses must work together to promote behavioral changes that support sustainable living. This includes integrating sustainability education into curricula, creating incentives for sustainable consumption, and using media and communication campaigns to raise awareness about the importance of the SDGs. By fostering a culture of sustainability, individuals and communities will be more likely to adopt environmentally and socially responsible practices.

The SDGs are ambitious, and achieving them within the set timeframe may seem impossible under the current trajectory. However, by adopting innovative financing mechanisms, strengthening global governance, leveraging technology, promoting circular economies, and empowering local communities, we can begin to turn the impossible into the possible. The road to achieving the SDGs will require bold action, collaboration, and commitment from all sectors of society. While the path forward is challenging, the potential rewards—a more just, equitable, and sustainable world—are worth the effort.

The future of sustainability principles and practices is one that promises a more just, resilient, and prosperous world for all. By embracing creativity, collaboration, and inclusivity, we can navigate the complexities of sustainability and create systems that honor both the environment and humanity. It is through this collective commitment to transforming our approaches to sustainability that we can ensure a thriving planet for future generations, where harmony between humanity and the natural world is not just an aspiration, but a reality.

References

Abanades, J. C., Rubin, E. S., Mazzotti, M., and Herzog, H. J. (2017). On the climate change mitigation potential of CO 2 conversion to fuels. *Energy and Environmental Science*, *10*(12), 2491–2499.

Ahrens, J. (2006). Governance in the process of economic transformation. *Private University of Applied Sciences Goettingen.*

Aitken, M., Haggett, C., and Rudolph, D. (2016). Practices and rationales of community engagement with wind farms: awareness raising, consultation, empowerment. *Planning Theory and Practice*, *17*(4), 557–576.

Anup, K. C., Bhandari, G., Joshi, G. R., and Aryal, S. (2013). Climate change mitigation potential from carbon sequestration of community forest in mid hill region of Nepal. *International Journal of Environmental Protection*, *3*(7), 33.

Arruda, E. H., Melatto, R. A. P. B., Levy, W., and de Melo Conti, D. (2021). Circular economy: A brief literature review (2015–2020). *Sustainable Operations and Computers*, *2*, 79–86.

Banga, J. (2019). The green bond market: a potential source of climate finance for developing countries. *Journal of Sustainable Finance and Investment*, *9*(1), 17–32.

Barrett, S. (1999). A theory of full international cooperation. *Journal of Theoretical Politics*, *11*(4), 519–541.

Bassi, S., and ten Brink, P. (2013). Step-by-step guidance: climate change mitigation (carbon storage and sequestration). *Social and Economic Benefits of Protected Areas: An Assessment Guide*, 156.

Behdadfar, E., and Samaei, S. R. (2024). Towards a Smart Tehran: Leveraging Machine Learning for Sustainable Development, Balanced Growth, and Resilience. *Journal of New Researches in Smart City*, *2*(2), 53–67.

Bhandari, M. (2024a). Citizen science and its applicability for sustainability and a healthy planet. *Academia Environmental Sciences and Sustainability*, *1*(1).

Bhandari, M. P. (2019). Institutional Goals of Sustainability in the Context of Higher Education-Contribution. *Zeszyty Naukowe Wy szej Szkoły Finansów i Prawa w Bielsku-Białej*, *23*(4), 5–12.

Bhandari, M. P. (2019). Sustainable development: Is this paradigm the remedy for all challenges? Does its goals capture the essence of real development and sustainability? With reference to discourses, creativeness, boundaries, and institutional architecture. SocioEconomic Challenges, 3 (4), 97–128.

Bhandari, M. P. (2020). Security and sustainability. *Zeszyty Naukowe Wyższej Szkoły Finansów i Prawa w Bielsku-Białej*, *24*(4), 5–8.

Bhandari, M. P. (2022). *Green Web-II: Standards and Perspectives from the IUCN Program/Policy Development in Environment Conservation Domain-with reference to India, Pakistan, Nepal, and Bangladesh.* River Publishers.

Bhandari, M. P. (2023). The fundamental principles of social sciences. *Business Ethics and Leadership*, *7*(2), 73–86.

Bhandari, M. P. (2024). Feminisms in Social Sciences. In *Women and Society.* IntechOpen.

Bhandari, M. P. (2024). *The Nexus of Climate Change and Land-use–Global Scenario with Reference to Nepal.* CRC Press.

Bhandari, M. P. (2024b). What is next for the sustainable development goals, what are the challenges concerning SDG 10—reduced inequalities?. *Sustainable Earth Reviews*, *7*(1), 23.

Bhandari, M. P., and Bhattarai, K. (2017). Institutional architecture for sustainable development (SD): A case study from Bangladesh, India, Nepal, and Pakistan.

Bhandari, M. P., and Hanna, S. (2022). Introduction: The Problems and Consequences of Sustainable Development Goals. In *Reducing Inequalities Towards Sustainable Development Goals* (pp. 1–25). River Publishers.

Bhandari, M.P. and Adhikari, M (2024). *Navigating Sustainability: A Comparative Analysis of Micro vs. Macro Strategies in Least Developed Countries*, Nova Science Publishers DOI: https://doi.org/10.52305/LNVZ4146, ISBN: 979-8-89113-771-4

Bhandari, P. (2024c). Book Review on Medani P. Bhandari (2023). Live and Let Others Live–In Reference to Sustainability and Environment Conservation. *Health Economics and Management Review*, *5*(2), 166–172.

Borón, V., Payán, E., MacMillan, D., and Tzanopoulos, J. (2016). Achieving sustainable development in rural areas in Colombia: Future scenarios for biodiversity conservation under land use change. *Land use policy*, *59*, 27–37.

Burnett, C., Swanberg, M., Hudson, A., and Schminkey, D. (2018). Structural Justice: A critical feminist framework exploring the intersection between justice, equity, and structural reconciliation. *Journal of Health Disparities Research and Practice*, *11*(4), 4.

Cahalan, M. W., Brunt, N., Vaughan III, T., Montenegro, E., Breen, S., Ruffin, E., and Perna, L. W. (2024). *Indicators of Higher Education Equity In the United States: 2024 50-Year Historical Trend Report. The Paradox of US Higher Education: Confronting Realities, Exploring Solutions.* Pell Institute for the Study of Opportunity in Higher Education.

Carpenter, S. R., Bennett, E. M., and Peterson, G. D. (2006). Scenarios for ecosystem services: an overview. *Ecology and Society*, *11*(1).

Cattapan, A., Acker-Verney, J. M., Dobrowolsky, A., Findlay, T., and Mandrona, A. (2020). Community engagement in a time of confinement. *Canadian Public Policy*, *46*(S3), S287-S299.

Cheba, K., Bąk, I., and Pietrzak, M. B. (2023). Conditions of the green transformation. The case of the European Union. *Technological and Economic Development of Economy*, *29*(2), 438–467.

Cheba, K., Bąk, I., Szopik-Depczyńska, K., and Ioppolo, G. (2022). Directions of green transformation of the European Union countries. *Ecological Indicators*, *136*, 108601.

Clow, M. (1998). The natural limits of technological innovation. *Technology in society*, *20*(2), 141–156.

Daba, M. H., and Dejene, S. W. (2018). The role of biodiversity and ecosystem services in carbon sequestration and its implication for climate change mitigation. *Environmental Sciences and Natural Resources*, *11*(2), 1–10.

Davelaar, E. J., and Nijkamp, P. (1990). Technological innovation and spatial transformation. *Technological Forecasting and Social Change*, *37*(2), 181–202.

De Vries, G., Arfelt, L., Drees, D., Godemann, M., Hamilton, C., Jessen-Thiesen, B., ... and Woltjer, P. (2021). The Economic Transformation Database (ETD): content, sources, and methods. *Technical Note, WIDER*.

Derichs, C. (2017). *Knowledge production, area studies, and global cooperation*. Routledge.

Dessai, S., and Hulme, M. (2004). Does climate adaptation policy need probabilities? *Climate policy*, *4*(2), 107–128.

Diaz, A., Schöggl, J. P., Reyes, T., and Baumgartner, R. J. (2021). Sustainable product development in a circular economy: Implications for products, actors, decision-making support and lifecycle information management. *Sustainable Production and Consumption*, *26*, 1031–1045.

Doan, Y. T., and Nguyen, H. T. (2023). Job opportunities of employees in the context of digital transformation. *World Journal of Advanced Research and Reviews*, *18*(3), 079–085.

Dolan, M. (1993). Global economic transformation and less developed countries. *Slater, etal (eds). Global Transformation and The Third World. London: Lynne Renner Publishers.*

Eden, C. A., Chisom, O. N., and Adeniyi, I. S. (2024). Education policy and social change: Examining the impact of reform initiatives on equity and access. *International Journal of Science and Research Archive, 11*(2), 139–146.

Edwards, M. G. (2021). The growth paradox, sustainable development, and business strategy. *Business Strategy and the Environment, 30*(7), 3079–3094.

Ekardt, F. (2019). *Sustainability: transformation, governance, ethics, law.* Springer.

Elmqvist, T., Setälä, H., Handel, S. N., van der Ploeg, S., Aronson, J., Blignaut, J. N., ... and de Groot, R. (2015). Benefits of restoring ecosystem services in urban areas. *Current opinion in environmental sustainability, 14*, 101–108.

Enarson, D. A. (2000). Tuberculosis: 12. Global disease and the role of international collaboration. *CMAJ, 162*(1), 57–61.

Enarson, D. A. (2000). Tuberculosis: 12. Global disease and the role of international collaboration. *CMAJ, 162*(1), 57–61.

Erhart, M., Ottova, V., Gaspar, T., Jericek, H., Schnohr, C., Alikasifoglu, M., ... and HBSC Positive Health Focus Group. (2009). Measuring mental health and well-being of school children in 15 European countries using the KIDSCREEN-10 Index. *International journal of public health, 54*, 160–166.

Fawzy, S., Osman, A. I., Doran, J., and Rooney, D. W. (2020). Strategies for mitigation of climate change: a review. *Environmental Chemistry Letters, 18*, 2069–2094.

Ferri, G., and Lipari, F. (2019). Sustainable Finance Management. *The Rise of Green Finance in Europe: Opportunities and Challenges for Issuers, Investors and Marketplaces*, 79–93.

Fischer, A., Holstead, K., Hendrickson, C. Y., Virkkula, O., and Prampolini, A. (2017). Community-led initiatives' everyday politics for sustainability–Conflicting rationalities and aspirations for change?. *Environment and Planning A, 49*(9), 1986–2006.

Frankham, R., Ballou, J. D., Dudash, M. R., Eldridge, M. D., Fenster, C. B., Lacy, R. C., ... and Ryder, O. A. (2012). Implications of different species concepts for conserving biodiversity. *Biological Conservation, 153*, 25–31.

Fraser, A., Tan, S., Lagarde, M., and Mays, N. (2018). Narratives of promise, narratives of caution: A review of the literature on Social Impact Bonds. *Social policy and administration*, *52*(1), 4–28.

Fromel, K., Kudlacek, M., Groffik, D., Svozil, Z., Simunek, A., and Garbaciak, W. (2017). Promoting healthy lifestyle and well-being in adolescents through outdoor physical activity. *International journal of environmental research and public health*, *14*(5), 533.

Galvin, R., and Healy, N. (2020). The Green New Deal in the United States: What it is and how to pay for it. *Energy Research and Social Science*, *67*, 101529.

Gandhi, G. (2015). Charting the evolution of approaches employed by the Global Alliance for Vaccines and Immunizations (GAVI) to address inequities in access to immunization: a systematic qualitative review of GAVI policies, strategies and resource allocation mechanisms through an equity lens (1999–2014). *BMC Public Health*, *15*, 1–35.

Garud, R., Nayyar, P. R., Shapira, Z., and Shapira, Z. B. (Eds.). (1997). *Technological innovation: Oversights and foresights*. Cambridge University Press.

Georghiou, L. (1998). Global cooperation in research. *Research Policy*, *27*(6), 611–626.

Gerstlberger, W., Praest Knudsen, M., and Stampe, I. (2014). Sustainable development strategies for product innovation and energy efficiency. *Business Strategy and the Environment*, *23*(2), 131–144.

Gilmour, H. (2012). Social participation and the health and well-being of Canadian seniors. *Health reports*, *23*(4), 23–32.

Greenberg, J., and Cohen, R. L. (Eds.). (2014). *Equity and justice in social behavior*. Academic Press.

Gregson, N., Crang, M., Fuller, S., and Holmes, H. (2015). Interrogating the circular economy: the moral economy of resource recovery in the EU. *Economy and society*, *44*(2), 218–243.

Gu, S., Xie, M., and Zhang, X. (2019). *Green transformation and development*. Palgrave Macmillan.

Güney, T. (2019). Renewable energy, non-renewable energy and sustainable development. *International Journal of Sustainable Development and World Ecology*, *26*(5), 389–397.

Gürçam, S. (2024). Paving the Way for Climate Resilience through Sustainable Urbanization: A Comparative Study. *Lectio Socialis*, *8*(1), 17–34.

Guy, M. E., and McCandless, S. A. (2020). *Achieving social equity: From problems to solutions*. Irvine, CA: Melvin and Leigh.

Gwiszcz, J. M. (2018). *Opening up transformation pathways for sustainable wellbeing: exploring the role of sustainability experiential learning as a capacity building mechanism for global ecological citizenship*. Arizona State University.

Harlan, T. (2021). Green development or greenwashing? A political ecology perspective on China's green Belt and Road. *Eurasian Geography and Economics*, *62*(2), 202–226.

Harrison, P. A., Berry, P. M., Simpson, G., Haslett, J. R., Blicharska, M., Bucur, M., ... and Turkelboom, F. (2014). Linkages between biodiversity attributes and ecosystem services: A systematic review. *Ecosystem services*, *9*, 191–203.

Hart, D. K. (1974). Social equity, justice, and the equitable administrator. *Public Administration Review*, *34*(1), 3–11.

Hautekiet, P., Saenen, N. D., Martens, D. S., Debay, M., Van der Heyden, J., Nawrot, T. S., and De Clercq, E. M. (2022). A healthy lifestyle is positively associated with mental health and well-being and core markers in ageing. *BMC medicine*, *20*(1), 328.

Hidalgo, D., Martín-Marroquín, J. M., and Corona, F. (2019). A multi-waste management concept as a basis towards a circular economy model. *Renewable and Sustainable Energy Reviews*, *111*, 481–489.

Johnson, N. J., and Svara, J. H. (Eds.). (2011). *Justice for all: Promoting social equity in public administration*. ME Sharpe.

Kelly, B. J., Lewin, T. J., Stain, H. J., Coleman, C., Fitzgerald, M., Perkins, D., ... and Beard, J. R. (2011). Determinants of mental health and well-being within rural and remote communities. *Social psychiatry and psychiatric epidemiology*, *46*, 1331–1342.

Khatibi, F. S., Dedekorkut-Howes, A., Howes, M., and Torabi, E. (2021). Can public awareness, knowledge and engagement improve climate change adaptation policies?. *Discover Sustainability*, *2*, 1–24.

Khoday, K., and Perch, L. (2012). Green Equity: Environmental Justice for More Inclusive Growth. *Policy Research Brief, International Policy Centre for Inclusive Growth, 19.*

Kirchherr, J., Reike, D., and Hekkert, M. (2017). Conceptualizing the circular economy: An analysis of 114 definitions. *Resources, conservation and recycling*, *127*, 221–232.

Kiss, A. (2004). Making biodiversity conservation a land-use priority. In *Getting biodiversity projects to work: Towards more effective conservation and development* (pp. 98–123). Columbia University Press.

Korpela, K. (2018). Can Rwanda leapfrog to the digital economy with ICT enabled development? A case study of a developing country determined to modernize without the traditional model.

Kuramoto, J., and Sagasti, F. (2002). Integrating local and global knowledge, technology and production systems: challenges for technical cooperation. *Science, Technology and Society*, *7*(2), 215–247.

Lay, J., and Tafese, T. (2020). *Promoting private investment to create jobs: A review of the evidence*. Kiel Institute for the World Economy (IfW), Poverty Reduction, Equity and Growth Network (PEGNet).

Lee, K. H. (2015). Drivers and barriers to energy efficiency management for sustainable development. *Sustainable Development*, *23*(1), 16–25.

Lee, S. M. (2003). South Korea: From the land of morning calm to ICT hotbed. *Academy of Management Perspectives*, *17*(2), 7–18.

Lee, V. H., Ooi, K. B., Chong, A. Y. L., and Seow, C. (2014). Creating technological innovation via green supply chain management: An empirical analysis. *Expert Systems with Applications*, *41*(16), 6983–6994.

Lewandowski, M. (2016). Designing the business models for circular economy—Towards the conceptual framework. *Sustainability*, *8*(1), 43.

Li, G., Wang, X., Su, S., and Su, Y. (2019). How green technological innovation ability influences enterprise competitiveness. *Technology in Society*, *59*, 101136.

Lipp, J. (2007). Lessons for effective renewable electricity policy from Denmark, Germany and the United Kingdom. *Energy policy*, *35*(11), 5481–5495.

Lund, H. (2007). Renewable energy strategies for sustainable development. *energy*, *32*(6), 912–919.

Malinauskaite, J., Jouhara, H., Czajczy ska, D., Stanchev, P., Katsou, E., Rostkowski, P., ... and Spencer, N. (2017). Municipal solid waste management and waste-to-energy in the context of a circular economy and energy recycling in Europe. *Energy*, *141*, 2013–2044.

Martín-María, N., Caballero, F. F., Moreno-Agostino, D., Olaya, B., Haro, J. M., Ayuso-Mateos, J. L., and Miret, M. (2020). Relationship between subjective well-being and healthy lifestyle behaviours in older adults: a longitudinal study. *Aging and mental health*, *24*(4), 611–619.

Martin-Ortega, J., Ferrier, R. C., Gordon, I. J., and Khan, S. (2015). *Water ecosystem services: a global perspective*. UNESCO Publishing.

Massetti, E., and Mendelsohn, R. (2018). Measuring climate adaptation: Methods and evidence. *Review of Environmental Economics and Policy*.

Matisoff, D. C., Noonan, D. S., and O'Brien, J. J. (2013). Convergence in environmental reporting: Assessing the carbon disclosure project. *Business Strategy and the Environment*, *22*(5), 285–305.

Matisoff, D. C., Noonan, D. S., and O'Brien, J. J. (2013). Convergence in environmental reporting: Assessing the carbon disclosure project. *Business Strategy and the Environment*, *22*(5), 285–305.

Mcdaid, D., and Park, A. L. (2011). Investing in mental health and well-being: findings from the DataPrev project. *Health promotion international*, *26*(suppl_1), i108-i139.

Mehmood, S., Zaman, K., Khan, S., and Ali, Z. (2024). The role of green industrial transformation in mitigating carbon emissions: Exploring the channels of technological innovation and environmental regulation. *Energy and Built Environment*, *5*(3), 464–479.

Mennillo, G., Schlenzig, T., and Friedrich, E. (Eds.). (2011). *Balanced Growth: finding strategies for sustainable development.* Springer Science and Business Media.

Mensah, C. N., Long, X., Dauda, L., Boamah, K. B., Salman, M., Appiah-Twum, F., and Tachie, A. K. (2019). Technological innovation and green growth in the Organization for Economic Cooperation and Development economies. *Journal of Cleaner Production*, *240*, 118204.

Mervelskemper, L., Kaltofen, D., and Stein, S. (2014). Are sustainable investment funds worth the effort?. *Journal of Sustainable Finance and Investment*, *4*(2), 127–146.

Mestre, A., and Cooper, T. (2017). Circular product design. A multiple loops life cycle design approach for the circular economy. *The Design Journal*, *20*(sup1), S1620-S1635.

Mikelsone, E., Atstaja, D., Koval, V., Uvarova, I., Mavlutova, I., and Kuzmina, J. (2021). Exploring sustainable urban transformation concepts for economic development. *Studies of Applied Economics*, *39*(5).

Miller, J. R., Groom, M., Hess, G. R., Steelman, T., Stokes, D. L., Thompson, J., ... and Marquardt, R. (2009). Biodiversity conservation in local planning. *Conservation Biology*, *23*(1), 53–63.

Miremadi, S. I. (2019). Technological innovation system: A scheme of innovation policy and technology development. *Journal of Science and Technology Policy*, *12*(2), 171–192.

Mundere, L., and Marešová, P. (2016). Potential of ICT Industry for Economic Growth in Developing Countries-A Case Study of the Republic of Rwanda. *Adv. Sci. Lett.*

Murray, M. G., Green, M. J. B., Bunting, G. C., and Paine, J. R. (1997). *Biodiversity conservation in the tropics: Gaps in habitat protection and funding priorities.* World conservation Press.

Nolan, J. (Ed.). (2010). *Global engagement: cooperation and security in the 21st century.* Brookings Institution Press.

Norbu, N. P., Tateno, Y., and Bolesta, A. (2021). Structural transformation and production linkages in Asia-Pacific least developed countries: An input-output analysis. *Structural Change and Economic Dynamics*, *59*, 510–524.

O'Mara-Eves, A., Brunton, G., Oliver, S., Kavanagh, J., Jamal, F., and Thomas, J. (2015). The effectiveness of community engagement in public health interventions for disadvantaged groups: a meta-analysis. *BMC Public Health*, *15*, 1-23.8.1.

Ochulor, O. J., Sofoluwe, O. O., Ukato, A., and Jambol, D. D. (2024). Technological innovations and optimized work methods in subsea maintenance and production. *Engineering Science and Technology Journal*, *5*(5), 1627–1642.

O'Mara-Eves, A., Brunton, G., McDaid, G., Oliver, S., Kavanagh, J., Jamal, F., ... and Thomas, J. (2013). Community engagement to reduce inequalities in health: a systematic review, meta-analysis and economic analysis. *Public Health Research*, *1*(4).

Osborne, C., Mayo, L., and Bussey, M. (2021). New frontiers in local government community engagement: Towards transformative place-based futures. *Futures*, *131*, 102768.

Østergaard, P. A., Duic, N., Noorollahi, Y., and Kalogirou, S. (2022). Renewable energy for sustainable development. *Renewable energy*, *199*, 1145–1152.

Otto, C., Reiss, F., Voss, C., Wüstner, A., Meyrose, A. K., Hölling, H., and Ravens-Sieberer, U. (2021). Mental health and well-being from childhood to adulthood: design, methods and results of the 11-year follow-up of the BELLA study. *European child and adolescent psychiatry*, *30*(10), 1559–1577.

Pagliusi, S., Che, Y., and Dong, S. (2019). The art of partnerships for vaccines. *Vaccine*, *37*(40), 5909–5919.

Paoli, G. D., Wells, P. L., Meijaard, E., Struebig, M. J., Marshall, A. J., Obidzinski, K., ... and d'Arcy, L. (2010). Biodiversity conservation in the REDD. *Carbon balance and management*, *5*, 1–9.

Paulo, S. (2014). International cooperation and development: A conceptual overview. *German Development Institute/Deutsches Institut für Entwicklungspolitik Discussion Paper*, *13*.

Pediatrics, *30*(4), 350–361.

Pegels, A., and Lütkenhorst, W. (2014). Is Germany's energy transition a case of successful green industrial policy? Contrasting wind and solar PV. *Energy Policy*, *74*, 522–534.

Peter, F. (2001). Health equity and social justice. *Journal of Applied Philosophy*, 159–170.

Petrillo, A., and Bellaviti, P. (2018). Sharing knowledge for change. Universities and new cultures of cooperation: transnational research and higher education for sustainable global urban development. *P. Agostino, and P. Bellaviti, Sustainable Urban Development and Globalization: New strategies for new challenges—with a focus on the Global South (Research for Development).*

Pranis, K. (2001). Restorative justice, social justice, and the empowerment of marginalized populations. *Restorative community justice: Repairing harm and transforming communities*, 287–306.

Quitzow, L., Canzler, W., Grundmann, P., Leibenath, M., Moss, T., and Rave, T. (2016). The German Energiewende–What's happening? Introducing the special issue. *Utilities Policy*, *41*, 163–171.

Raman, VijayaVenkata, S., Iniyan, S., and Goic, R. (2012). A review of climate change, mitigation and adaptation. *Renewable and Sustainable Energy Reviews*, *16*(1), 878–897.

Rawat, U., and Agarwal, N. K. (2015). Biodiversity: Concept, threats, and conservation. *Environment Conservation Journal*, *16*(3), 19–28.

Rosenberg, M. L. (2010). *Real collaboration: what it takes for global health to succeed* (Vol. 20). Univ of California Press.

Rosenberg, M. L. (2010). *Real collaboration: what it takes for global health to succeed* (Vol. 20). Univ of California Press.

Salikhova, O. B., and Krekhivsky, O. V. (2018). A New Mechanism for State Support to Technological Innovation for Industrial Development. *Statistics of Ukraine*, *81*(2), 30–35.

Sepetis, A., Pierrakos, G., and Schallmo, D. (2022). Sustainable Finance and the Digital Transformation in the Healthcare system. In *ISPIM Conference Proceedings* (pp. 1–21). The International Society for Professional Innovation Management (ISPIM).

Shaeffer, S. (2019). Inclusive education: a prerequisite for equity and social justice. *Asia Pacific Education Review*, *20*(2), 181–192.

Shah, S. H., Rodina, L., Burt, J. M., Gregr, E. J., Chapman, M., Williams, S., ... and McDowell, G. (2018). Unpacking social-ecological transformations: Conceptual, ethical and methodological insights. *The Anthropocene Review*, *5*(3), 250–265.

Shalowitz, M. U., Isacco, A., Barquin, N., Clark-Kauffman, E., Delger, P., Nelson, D., ... and Wagenaar, K. A. (2009). Community-based participatory research: a review of the literature with strategies for community engagement. *Journal of Developmental and Behavioral Pediatrics*, *30*(4), 350–361.

Shi, H., Singh, A., Kant, S., Zhu, Z., and Waller, E. (2005). Integrating habitat status, human population pressure, and protection status into biodiversity conservation priority setting. *Conservation biology*, 1273–1285.

Shi, L., and Moser, S. (2021). Transformative climate adaptation in the United States: Trends and prospects. *Science*, *372*(6549), eabc8054.

Steil, B., Victor, D. G., and Nelson, R. R. (Eds.). (2002). *Technological innovation and economic performance*. Princeton University Press.

Suarez, V. (2024). Community engagement: exploring participatory territorial approaches in municipal culture agendas. *Journal of Social Sciences and Humanities Research Fundamentals*, *4*(04), 47–52.

Subramoniam, R., Sundin, E., Subramoniam, S., and Huisingh, D. (2021). Riding the digital product life cycle waves towards a circular economy. *Sustainability*, *13*(16), 8960.

Sumter, D., de Koning, J., Bakker, C., and Balkenende, R. (2020). Circular economy competencies for design. *Sustainability*, *12*(4), 1561.

Tomić, T., and Schneider, D. R. (2020). Circular economy in waste management–Socio-economic effect of changes in waste management system structure. *Journal of Environmental Management*, *267*, 110564.

Topp, C. W., Østergaard, S. D., Søndergaard, S., and Bech, P. (2015). The WHO-5 Well-Being Index: a systematic review of the literature. *Psychotherapy and psychosomatics*, *84*(3), 167–176.

Vuorisalo, T., and Laihonen, P. (2000, January). Biodiversity conservation in the north: history of habitat and species protection in Finland. In *Annales Zoologici Fennici* (pp. 281–297). Finnish Zoological and Botanical Publishing Board.

Wagner, C. S., Park, H. W., and Leydesdorff, L. (2015). The continuing growth of global cooperation networks in research: A conundrum for national governments. *PloS one*, *10*(7), e0131816.

Waham, J. J., Asfahani, A., and Ulfa, R. A. (2023). International Collaboration in Higher Education: Challenges and Opportunities in a Globalized World. *EDUJAVARE: International Journal of Educational Research*, *1*(1), 49–60.

Wainaina, S., Awasthi, M. K., Sarsaiya, S., Chen, H., Singh, E., Kumar, A., ... and Taherzadeh, M. J. (2020). Resource recovery and circular economy from organic solid waste using aerobic and anaerobic digestion technologies. *Bioresource Technology*, *301*, 122778.

Winterstetter, A., Heuss-Assbichler, S., Stegemann, J., Kral, U., Wäger, P., Osmani, M., and Rechberger, H. (2021). The role of anthropogenic resource classification supports the transition to a circular economy. *Journal of Cleaner Production*, *297*, 126753.

Woodward, R. (2007). Global governance and the Organization for Economic Cooperation and Development. In *Global Governance and Japan* (pp. 83–99). Routledge

Young, J., Watt, A., Nowicki, P., Alard, D., Clitherow, J., Henle, K., ... and Richards, C. (2005). Towards sustainable land use: Identifying and managing the conflicts between human activities and European biodiversity conservation. *Biodiversity and Conservation*, *14*, 1641–1661.

Zakari, A., Khan, I., Tan, D., Alvarado, R., and Dagar, V. (2022). Energy efficiency and sustainable development goals (SDGs). *Energy*, *239*, 122365.

Index

S

Words / Phrase Definitions

Actionable insights: Clear, practical information derived from research or analysis that can be used to inform decisions and guide action. In environmentalism, actionable insights help policymakers, businesses, and communities implement effective sustainability practices.

Balanced growth: Economic growth that considers environmental and social impacts, aiming for long-term sustainability rather than short-term profits. Balanced growth promotes equitable distribution of wealth, resource efficiency, and ecosystem health.

Behavioral change: The process of altering individuals' habits, attitudes, or actions to promote more sustainable or health-conscious lifestyles. Behavioral change initiatives often target areas like reducing energy consumption, adopting healthier eating habits, or shifting toward eco-friendly products. These efforts can be driven by education, policies, incentives, and cultural shifts.

Biodiversity conservation: The protection and management of biodiversity to ensure the survival of species and ecosystems. Biodiversity conservation is crucial for maintaining ecosystem services, food security, and the resilience of natural systems.

Capacity building: Efforts to enhance the skills, knowledge, and resources of individuals, organizations, or communities, enabling them to effectively address challenges and pursue development. Capacity building empowers local actors to manage projects, drive change, and ensure long-term success.

Carbon sequestration: The process of capturing and storing atmospheric carbon dioxide in plants, soils, oceans, or artificial structures. Carbon sequestration helps mitigate climate change by reducing the concentration of CO2 in the atmosphere.

Circular economy: A circular economy is an economic system that prioritizes the reuse, recycling, and regeneration of resources to minimize waste and reduce environmental impact. Unlike the traditional linear economy, which follows a "take, make, dispose" model, the circular economy aims to close the loop by designing products for longevity, repairability, and recyclability. This system encourages sustainable production and consumption, reducing resource extraction and waste generation, and promotes the efficient use of resources. Example: In a circular economy, businesses focus on using biodegradable materials for packaging, creating products that can be easily disassembled for recycling, and developing services such as product leasing to extend the life cycle of products.

Circular supply chain: A model in which products and materials are designed, produced, and managed in a way that minimizes waste and maximizes reuse, recycling, and regeneration. Circular supply chains aim to create a closed-loop system, reducing the need for virgin materials and lowering environmental impacts.

Climate adaptation: Adjusting social, economic, and environmental practices to minimize the negative impacts of climate change. Climate adaptation strategies include building resilient infrastructure, developing drought-resistant crops, and enhancing disaster preparedness.

Climate change mitigation: Actions taken to reduce or prevent the emission of greenhouse gases, thereby limiting the severity of future climate change. Mitigation strategies include transitioning to renewable energy, improving energy efficiency, and protecting carbon sinks like forests.

Climate change: The long-term alteration of temperature and weather patterns, largely due to human activities like burning fossil fuels. Climate change poses serious threats to ecosystems, human health, food security, and global stability.

Climate resilience: The ability of systems—whether natural, social, or economic—to withstand, recover from, and adapt to the impacts of climate change. Building climate resilience involves improving infrastructure, fostering adaptive capacities in communities, and developing policies that address both immediate and long-term climate risks, such as extreme weather events, rising sea levels, or changing agricultural conditions.

Collaborative efforts: The coordinated and cooperative actions of multiple stakeholders—such as governments, businesses, NGOs, and communities—working together toward a shared objective. Effective collaboration typically involves pooling resources, expertise, and ideas to tackle complex challenges, fostering innovation, improving outcomes, and building trust among participants. These efforts are essential in addressing global issues like climate change, social inequality, and sustainable development.

Community engagement: The process by which local populations are actively involved in the planning, decision-making, and implementation of projects or policies that affect them. True community engagement fosters a sense of ownership, improves the relevance and sustainability of initiatives, and empowers marginalized groups to have a voice in issues such as local governance, urban planning, and environmental conservation.

Corporate responsibility: The ethical obligation of businesses to contribute to social, environmental, and economic well-being beyond profit-making. Corporate responsibility involves adopting sustainable practices, supporting communities, promoting fair labor standards, reducing environmental impacts, and ensuring ethical supply chains. Companies that embrace corporate responsibility often enhance their brand reputation and customer loyalty.

Data gaps: Missing or incomplete data that limit understanding and decision-making in areas like climate change, biodiversity loss, or pollution. Addressing data gaps through better monitoring and research is crucial for informed environmental management.

Digital transformation: The integration of digital technologies into various aspects of society, driving innovation, efficiency, and sustainability. In the context of sustainability, digital transformation can enhance monitoring, data collection, and management systems for environmental protection.

Economic systems: The structures and processes that govern production, distribution, and consumption of goods and services in a society. Sustainable economic systems prioritize resource efficiency, equity, and environmental protection, contrasting with traditional models focused on growth at any cost.

Economic transformation: A significant restructuring of an economy's sectors, industries, and institutions to promote development, inclusion, and sustainability. Economic transformation may involve transitioning from resource-based or low-income economies to diversified, high-value sectors such as technology, services, and green industries. It aims to improve living standards, create jobs, and foster long-term resilience.

Ecosystem protection: Safeguarding natural environments and the biodiversity they support by implementing conservation practices, policies, and sustainable land use strategies. Ecosystem protection is essential for maintaining the services ecosystems provide, such as clean air, water, and pollination, which are critical for human survival and well-being.

Ecosystem services: The benefits that humans derive from natural ecosystems, including provisioning services (such as food and water), regulating services (like climate regulation and flood control), cultural services (recreational and spiritual benefits), and supporting services (like nutrient cycling and soil formation). Ecosystem services are critical for human survival and well-being, as they sustain natural processes and contribute to economic, social, and environmental health.

Ecosystem-based management: An integrated approach to managing natural resources that considers the entire ecosystem, including human activities, rather than focusing on individual species or resources. Ecosystem-based management seeks to balance ecological, social, and economic goals while maintaining the health of ecosystems over the long term. This approach is often used in fisheries, forestry, and conservation.

Education and empowerment: Providing individuals with the knowledge, skills, and confidence to take action on issues that affect their lives. Education and empowerment in sustainability help people understand environmental challenges and become agents of change in their communities.

Emission reductions: The decrease in the release of pollutants, particularly greenhouse gases, into the atmosphere. Achieving significant emission reductions is essential to combat climate change and improve air quality.

Energy efficiency: The use of technology and practices that reduce the amount of energy required to provide goods and services. Improving energy efficiency lowers costs, reduces emissions, and conserves resources, contributing to sustainability

Environmental degradation: The deterioration of the natural environment through depletion of resources, destruction of ecosystems, and pollution. Environmental degradation is driven by factors such as industrial activities, deforestation, and urbanization.

Environmental discourse: The dialogue and exchange of ideas concerning environmental issues. Environmental discourse shapes public understanding, policies, and actions related to climate change, conservation, and sustainability.

Environmental justice: The fair distribution of environmental benefits and burdens across all segments of society, regardless of race, class, or geography. Environmental justice seeks to ensure that marginalized communities are not disproportionately affected by environmental hazards, such as pollution or industrial development, and that they have equitable access to clean air, water, and natural resources.

Environmental sustainability: The responsible use and management of natural resources to prevent degradation and ensure the long-term health of ecosystems. It involves reducing pollution, conserving biodiversity, and addressing climate change. Environmental sustainability is a core principle in shaping policies that protect natural resources while supporting economic and social well-being.

Environmentalism transformation: The evolution of environmentalism from grassroots advocacy and conservation efforts into a more comprehensive, systemic approach that addresses global environmental crises. It includes shifts in policy, corporate responsibility, and individual behavior to promote sustainability, biodiversity conservation, and climate resilience.

ESG integration: The incorporation of environmental, social, and governance factors into investment decision-making processes. ESG integration allows investors to assess risks and opportunities beyond financial performance, leading to more sustainable and ethical investment practices.

Financial commitment: The allocation of financial resources to support sustainability initiatives, policies, or projects. Financial commitment from governments, businesses, and international organizations is essential to funding climate action, conservation efforts, and the transition to a green economy.

Global collaboration: International cooperation between countries, organizations, and institutions to address global challenges, such as climate change, poverty, and pandemics. Global collaboration often involves knowledge exchange, policy alignment, and coordinated efforts to develop solutions that transcend national borders, fostering peace, sustainability, and shared prosperity.

Global cooperation: Global cooperation involves collaboration between countries, international organizations, and other global stakeholders to address challenges that transcend national borders, such as climate change, biodiversity loss, and poverty. In the context of sustainability, global cooperation is essential for creating unified responses to environmental crises, ensuring equitable access to resources, and fostering peace and prosperity on a global scale. Example: The Paris Agreement is a result of global cooperation, where countries agreed to collectively limit global warming to well below 2°C and pursue efforts to limit the temperature increase to 1.5°C above pre-industrial levels.

Global sustainability: The collective effort of nations and organizations worldwide to achieve long-term ecological balance, social equity, and economic stability. Global sustainability requires cooperation on issues like climate change, resource management, and poverty alleviation.

Governance: The systems, processes, and institutions that guide decision-making and control within a society. Good governance in sustainability includes transparency, accountability, and the active participation of all stakeholders in addressing environmental challenges.

Green bonds: Debt instruments issued by governments, institutions, or corporations to finance projects that have positive environmental or climate benefits, such as renewable energy or pollution control. Green bonds offer investors the opportunity to support sustainable initiatives while earning returns.

Green economy: An economic system that focuses on reducing environmental risks and ecological scarcities while enhancing human well-being and social equity. The green economy promotes low-carbon, resource-efficient, and socially inclusive growth, supporting industries like renewable energy, sustainable agriculture, and environmental protection.

Green jobs: Employment opportunities in sectors that contribute to preserving or restoring the environment, such as renewable energy, energy efficiency, conservation, and sustainable agriculture. Green jobs support economic growth while addressing environmental challenges.

Green technologies: Innovations that reduce environmental impact by improving energy efficiency, minimizing waste, and utilizing renewable resources. Green technologies include solar energy, electric vehicles, and water-efficient systems, playing a vital role in achieving sustainability goals.

Green technology: Technologies that promote sustainability by reducing environmental impacts, conserving resources, or enhancing energy efficiency. Examples include solar panels, electric vehicles, and water purification systems.

Habitat protection: Efforts to safeguard natural environments from degradation, ensuring that species have the space and resources they need to survive. Protecting habitats helps maintain biodiversity and ecosystem health.

Habitat restoration: The process of returning a damaged or degraded natural environment to its original condition, often by reintroducing native species, restoring natural hydrology, or removing invasive species. Habitat restoration aims to rebuild ecosystems to support biodiversity, ecosystem services, and resilience against environmental threats such as climate change or habitat destruction.

Healthcare access: The ability of individuals to obtain necessary health services, including prevention, treatment, and rehabilitation. In sustainable development, healthcare access is linked to ensuring healthy lives and promoting well-being for all.

Human capital development: The process of improving individuals' skills, education, health, and abilities to enhance their economic and social contributions. Human capital development is a fundamental component of economic growth and social progress, as it increases productivity, innovation, and societal well-being. Investment in education, training, and healthcare systems are key aspects of this development.

Impact investing: Investments made with the intention of generating both financial returns and measurable social or environmental impacts. Impact investing supports initiatives such as affordable housing, education, renewable energy, and healthcare in underprivileged areas.

Inclusive growth: Economic growth that benefits all members of society, particularly marginalized or disadvantaged groups. It ensures that wealth and opportunities are equitably distributed, fostering social cohesion and reducing poverty and inequality. Inclusive growth often focuses on access to education, healthcare, jobs, and housing for vulnerable populations.

Inclusive policies: Policies designed to ensure that all individuals, especially marginalized and disadvantaged groups, have access to opportunities, resources, and decision-making processes. Inclusive policies in sustainability aim to promote equity and participation in addressing environmental and social issues.

Industrial diversification: The process of expanding an economy by developing a range of industries, rather than relying on a single sector. This reduces economic vulnerability and improves resilience against sector-specific downturns. Industrial diversification is critical for economic stability, particularly in regions dependent on agriculture, mining, or fossil fuels.

Infrastructure investment: The allocation of resources to build and maintain essential physical systems, such as transportation networks, energy grids, water supply, and communication systems. Investing in infrastructure enhances economic growth, improves quality of life, and facilitates access to basic services like healthcare, education, and clean water. Sustainable infrastructure investment also prioritizes resilience against climate change.

Innovation and technology: The creation and application of new ideas, processes, or products, often leveraging advanced technologies to solve problems or improve efficiency. Innovation drives economic growth and can address societal challenges, such as improving healthcare, education, and environmental sustainability. Technology plays a key role in enhancing productivity, connectivity, and access to resources.

Institutional reforms: Changes made to the structure, policies, or regulations of institutions to improve their efficiency, transparency, accountability, or fairness. Institutional reforms are crucial for addressing corruption, improving public service delivery, fostering democracy, and ensuring justice. Reforms may involve governance structures, legal systems, or public administration.

International agreements: International agreements are legally binding treaties or accords between countries to cooperate on global issues like environmental protection, trade, security, or human rights. In sustainability, international agreements are essential for coordinating actions on climate change, biodiversity conservation, pollution control, and resource management. These agreements create frameworks for cooperation, set common goals, and hold nations accountable for their environmental commitments. Example: The Kyoto Protocol and the Paris Agreement are international agreements aimed at reducing greenhouse gas emissions to mitigate climate change.

Job creation: The generation of new employment opportunities through economic growth, government policies, or business expansion. Job creation is vital for reducing poverty, improving living standards, and promoting social stability.

Knowledge gaps: Areas where information or understanding is incomplete or lacking, hindering effective decision-making or problem-solving. In sustainability, knowledge gaps might include insufficient data on climate change impacts, biodiversity loss, or the long-term effects of environmental policies.

Knowledge sharing: The exchange of information, skills, and expertise between individuals, organizations, or communities to improve decision-making, innovation, and problem-solving. Knowledge sharing promotes collective learning and progress, especially in areas such as sustainable development, education, and science, by disseminating best practices and fostering collaboration across sectors and regions.

Lifelong learning: The continuous development of knowledge and skills throughout an individual's life. In the context of sustainability, lifelong learning involves educating people about environmental issues, fostering adaptability to changing conditions, and encouraging sustainable behavior.

Local renewable energy: Energy generated from renewable sources, such as solar, wind, or biomass, within a specific community or region. Local renewable energy projects reduce dependence on fossil fuels, lower carbon emissions, and enhance energy security, often involving decentralized systems that empower communities and create local jobs.

Low-carbon development: Economic growth strategies that prioritize reducing greenhouse gas emissions, transitioning away from fossil fuels, and embracing clean energy technologies. Low-carbon development is critical for mitigating climate change while fostering economic resilience and sustainable growth.

Mental health: A state of well-being in which individuals realize their potential, cope with the normal stresses of life, work productively, and contribute to their community. Mental health encompasses emotional, psychological, and social well-being. It is crucial for overall health, affecting decision-making, relationships, and the ability to handle challenges. Addressing mental health involves promoting access to care, reducing stigma, and fostering supportive environments.

Monitoring and evaluation: Systematic processes used to assess the progress, performance, and outcomes of programs, policies, or projects. Monitoring involves tracking activities and outputs, while evaluation assesses their effectiveness and impact. These processes provide valuable insights for improving strategies and ensuring accountability and transparency.

Participatory governance: A democratic process in which citizens are actively involved in the governance and decision-making processes at local, national, or global levels. Participatory governance promotes transparency, accountability, and inclusiveness, empowering citizens to influence policies and decisions that affect their lives. Mechanisms include public consultations, community-led initiatives, and citizen councils.

Policy and governance: The frameworks, institutions, and processes that guide decision-making in public and private sectors. Effective governance involves creating policies that address societal challenges—such as environmental sustainability, social justice, and economic development—while ensuring transparency, accountability, and inclusiveness.

Policy formulation: The process of developing strategies, guidelines, and regulations to address public problems or challenges. It involves identifying issues, setting objectives, and crafting laws or rules to guide government action. Effective policy formulation requires research, stakeholder input, and alignment with societal values and goals, ensuring the policies are comprehensive and sustainable.

Policy frameworks: Structured guidelines that provide a foundation for decision-making in specific areas, such as environmental protection, public health, or economic development. Policy frameworks outline objectives, strategies, and mechanisms to achieve desired outcomes and ensure consistency and coherence in governance.

Political will: The determination of political leaders to pursue and implement policies that address pressing issues, such as climate change or social inequality. Strong political will is essential for enacting meaningful environmental policies and achieving sustainability goals.

Practical innovations: Applied solutions or advancements in methods and technologies that address real-world challenges. In environmentalism, practical innovations could involve new ways to reduce carbon emissions, conserve biodiversity, or manage natural resources sustainably.

Practical problems: Real-world obstacles that hinder the implementation of solutions or strategies. In environmentalism, practical problems may include insufficient funding for conservation, lack of political support, or technological limitations in renewable energy adoption.

Public awareness: Increasing the general public's knowledge and understanding of important issues, such as health, environmental conservation, or social justice. This can be achieved through media campaigns, educational programs, or grassroots movements. Effective public awareness helps shape attitudes and behaviors, encouraging more informed decisions and participation in societal issues.

Public-private partnerships: Collaborative agreements between government and private sector entities to finance, build, and operate projects that serve the public interest, such as transportation, energy, and social services. These partnerships leverage the expertise and efficiency of the private sector while serving public needs, often leading to more innovative and cost-effective solutions.

Regenerative agriculture: A farming approach that restores and enhances the health of ecosystems by prioritizing soil health, biodiversity, water cycles, and carbon sequestration. Unlike conventional agriculture, regenerative practices aim to rebuild soil organic matter, increase biodiversity, and reduce reliance on synthetic inputs, creating resilient, productive systems that also mitigate climate change.

Renewable energy: Energy derived from natural sources that are replenished on a human timescale, such as solar, wind, and hydropower. Transitioning to renewable energy is key to reducing greenhouse gas emissions and mitigating climate change.

Research and development (R&D): The process of developing new technologies, knowledge, or products through scientific research and experimentation. R&D is crucial for driving innovations that can address environmental challenges, such as renewable energy and climate adaptation technologies.

Resource efficiency: The optimal use of resources to minimize environmental impact while maximizing productivity and sustainability. Resource efficiency helps reduce waste, lower emissions, and conserve natural resources, such as water and energy.

Resource inequality: The unequal distribution of natural resources, wealth, and opportunities across different regions or populations. Resource inequality often leads to social and economic disparities, environmental degradation, and conflict over access to resources like water, land, and energy.

Resource recycling: Resource recycling involves the process of collecting, processing, and repurposing materials that would otherwise be considered waste. Recycling reduces the need for raw material extraction, lowers energy consumption, and minimizes environmental pollution. It plays a crucial role in promoting sustainability by transforming waste into valuable materials that can be reintroduced into the production cycle. Example: Recycling paper reduces the need for logging trees, and recycling metals like aluminum and steel saves significant amounts of energy compared to producing them from raw ores.

Shared knowledge: Shared knowledge refers to the exchange and dissemination of information, research, best practices, and technologies across different stakeholders, including governments, businesses, academia, and civil society. Sharing knowledge fosters collaboration, innovation, and capacity-building, helping communities, organizations, and countries make informed decisions and take effective actions toward sustainability. Example: International scientific collaborations, such as the Intergovernmental Panel on Climate Change (IPCC), play a key role in generating and sharing knowledge about climate science, which informs global climate policies and initiatives.

Skill development: Enhancing individuals' capabilities through education, training, and experience to improve their employability and productivity. Skill development is crucial for adapting to technological advances and changing labor market demands, fostering economic competitiveness and personal growth. Lifelong learning plays a vital role in ensuring workers can continuously update their skills.

SMEs support: Efforts to assist small and medium-sized enterprises (SMEs) through financial aid, training, regulatory simplifications, and market access. SMEs are crucial for job creation and innovation in many economies, so providing support helps stimulate economic growth, resilience, and competitiveness, particularly in emerging markets.

Social equity: The fair and just distribution of resources, opportunities, and privileges within society. Social equity in sustainability involves ensuring that marginalized communities have access to environmental benefits and are not disproportionately affected by environmental harms.

Social inclusion: Ensuring that all individuals, regardless of background or identity, have equal access to opportunities, resources, and services. Social inclusion promotes equity, reduces discrimination, and fosters cohesive, resilient communities by enabling marginalized groups to participate fully in society.

Social justice: A concept that calls for the fair treatment of all individuals within a society, addressing issues of inequality, discrimination, and oppression. In environmentalism, social justice emphasizes that vulnerable populations should not bear the brunt of environmental degradation.

Species conservation: The protection and management of species to prevent extinction, ensure population recovery, and maintain biodiversity. Conservation efforts focus on both endangered species and keystone species that play critical roles in ecosystems.

Supply chain transparency: The ability to trace and verify the origins, practices, and impacts of every stage of a product's life cycle—from raw materials to manufacturing and delivery. Supply chain transparency ensures that businesses and consumers are aware of the social, environmental, and ethical implications of their products, promoting accountability and responsible sourcing.

Sustainability concept: The principle of balancing environmental, social, and economic considerations to ensure long-term well-being for people and the planet. It emphasizes interdependence between human systems and natural ecosystems.

Sustainability culture: A societal mindset that prioritizes the long-term health of the planet and future generations in decision-making and everyday actions. Sustainability culture is rooted in values of conservation, social equity, and environmental stewardship, fostering collective responsibility for addressing issues like climate change, resource depletion, and biodiversity loss.

Sustainable Development Goals (SDGs): A set of 17 global goals adopted by the United Nations in 2015 to address a wide range of global challenges, including poverty, inequality, environmental degradation, and climate change, with the goal of achieving a more sustainable and equitable world by 2030.

Sustainable Development: Development that meets the needs of the present without compromising the ability of future generations to meet their own needs. It integrates environmental protection, economic growth, and social equity, aiming to balance progress with ecological preservation.

Sustainable finance: Investment approaches that consider environmental, social, and governance (ESG) factors alongside financial returns. Sustainable finance directs capital toward projects and businesses that promote sustainability, such as renewable energy, green infrastructure, and social inclusion.

Sustainable future: A vision of the world in which economic, social, and environmental needs are met without compromising the ability of future generations to meet theirs. This concept includes the responsible use of natural resources, reducing pollution, promoting equity, and ensuring long-term ecosystem health.

Sustainable land use: Managing land resources in a way that meets human needs while preserving ecosystem health. Sustainable land use practices include responsible agriculture, urban planning, and conservation efforts to prevent deforestation and soil degradation.

Sustainable practices: Methods of operating that meet current needs without compromising the ability of future generations to meet theirs. This includes practices in agriculture, business, and daily life that minimize resource consumption, reduce environmental harm, and promote social equity. Examples include using renewable energy, reducing waste, and supporting fair labor practices.

Sustainable waste management: The responsible collection, processing, recycling, and disposal of waste materials in a manner that minimizes harm to the environment and conserves resources. Sustainable waste management promotes waste reduction, reuse, and recycling, and aims to reduce landfill use, greenhouse gas emissions, and environmental pollution.

Technological gaps: The disparity between current technological capabilities and the innovations needed to achieve sustainability goals. Technological gaps may hinder progress in areas like renewable energy, emissions reduction, and climate adaptation.

Technological innovation: The development and application of new technologies to solve problems or improve processes. Technological innovation drives economic growth, improves efficiency, and can help address global challenges such as climate change, healthcare, and education. It includes advancements in fields like artificial intelligence, renewable energy, biotechnology, and communication technologies.

Theoretical innovations: New ideas or models that advance understanding within a field. In sustainability and environmentalism, theoretical innovations might include new frameworks for understanding human-environment interactions or rethinking economic systems in ways that promote sustainability.

Theoretical problems: Challenges in understanding or explaining a concept or phenomenon within a theoretical framework. In sustainability, theoretical problems may arise in defining the relationship between economic growth and ecological limits or conceptualizing justice in environmental policies.

Trade and investment: The exchange of goods, services, and capital between countries or regions, which drives economic growth and development. Trade enables countries to specialize in goods they produce efficiently, while investment supports the development of new industries and infrastructure. Promoting fair trade and sustainable investment is key to reducing global inequalities.

Transformative education: An approach to education that empowers learners to critically analyze their world, challenge inequities, and contribute to sustainable and just social transformations. In environmentalism, transformative education fosters deep understanding of ecological and social systems and inspires action for sustainability.

Waste management: Waste management refers to the collection, transportation, treatment, and disposal of waste materials in a way that minimizes their impact on human health and the environment. Effective waste management includes reducing waste generation, promoting recycling and composting, and managing landfills and incineration. Sustainable waste management also focuses on minimizing hazardous waste and finding alternatives to disposal, such as energy recovery and circular economy practices. Example: Many cities are adopting zero-waste strategies, promoting composting of organic waste, recycling of materials like glass and plastics, and using waste-to-energy technologies to convert waste into electricity or heat.

About the Author

Dr. Medani P. Bhandari is a widely recognized humanitarian, professor, author, and editor who has significantly contributed to social and environmental sciences. With expertise in social equity, inequality, inclusion, innovation, feminism, sustainability, climate change, and social and environmental policies and management, Dr. Bhandari has dedicated his career to these important areas. In addition to his extensive scholarly work, Dr. Bhandari is a poet, essayist, and advocate for the environment and social causes. His expertise expands to global and international environmental politics and justice, climate change, sustainable development, public/social policy, the non-profit sector, renewable energy, nature, culture, and power. He has gained field experience across diverse regions, including Asia, Africa, North America, Western Europe, Australia, Japan, and the Middle East. Dr. Bhandari has authored and co-edited several books, including Green Web-II: Standards and Perspectives from the IUCN, 2nd Edition, Getting the Climate Science Facts Right: The Role of the IPCC, Reducing Inequalities Towards Sustainable Development Goals: Multilevel Approach, Educational Transformation, Economic Inequality – Trends, Traps, and Trade-offs, Social Inequality as a Global Challenge, The Unbeatable Challenges of Inequality, Perspectives on Sociological Theories, Methodological Debates, and Organizational Sociology, The Past, Present, and Future Directions of Migration/Immigration, Theories, and Practices of Global Journeys, Transforming Futures – Navigating Society, Environment, and Governance, Live and Let Others Live – In Reference to Sustainability and Environment Conservation, Navigating Sustainability: A Comparative Analysis of Micro vs. Macro Strategies in the least Developed Countries, The Nexus of Climate Change and Land-use – Global Scenario with Reference to Nepal, Women and Society etc.. Social Inequality- Past, Present, and Future (in which 15 African Countries' case studies are included), etc. Furthermore, Dr. Bhandari's creative writing encompasses a vast collection of poems and essays. He has published three volumes of poetry in collaboration with Prajita Bhandari. Dr. Bhandari's multifaceted contributions to academia, humanitarianism, and creative writing have solidified his reputation as an esteemed figure in social and environmental sciences, profoundly impacting social equity, sustainability, climate change, and beyond. Professor Bhandari holds five master's degrees in social and environmental sciences and a PhD in sociology. For 45 years, he has been actively involved in environmental conservation and humanitarian efforts through nonprofit and academic platforms. Guided by the philosophy that all living and non-living beings are interconnected ("Vasudhaiva Kutumbakam") and the principle of "Live and Let Others Live," his work reflects a deep commitment to the well-being of our planet and its inhabitants.

He serves as an International Program Director at the New York-based NGO Atlantic State Legal Foundation in Syracuse, NY; a professor at Akamai University, USA, and Sumy State University, Ukraine; and an advisor to Gandaki University, Nepal. He is also the UNEP Civil Society Regional Facilitator for North America and the Lead Author of UNEP GEO 7. He is also the editor-in-chief of Strategic Planning for Energy and the Environment, the books series editor at the River Publisher, and the books series editor at Nova Publishers.

For Product Safety Concerns and Information please contact our EU representative GPSR@taylorandfrancis.com
Taylor & Francis Verlag GmbH, Kaufingerstraße 24, 80331 München, Germany

www.ingramcontent.com/pod-product-compliance
Ingram Content Group UK Ltd.
Pitfield, Milton Keynes, MK11 3LW, UK
UKHW022223190926
475702UK00003B/485

* 9 7 8 8 7 7 0 0 4 8 1 2 5 *